T0230685

# Physical Sensors for Biomedical Applications

Editors:

**Michael R. Neuman, Ph.D, M.D.**
Associate Professor of Biomedical
Engineering in Reproductive Biology
Department of Reproductive Biology
Case Western Reserve University
School of Medicine
Cleveland, Ohio

**David G. Fleming, Ph.D., M.D.**
Professor of Biomedical Engineering
Department of Pediatrics
Case Western Reserve University and
Rainbow Babies and Childrens Hospital
Cleveland, Ohio

**Peter W. Cheung, Ph.D.**
Associate Professor
Department of Biomedical Engineering
Case Western Reserve University
Cleveland, Ohio

**Wen H. Ko, Ph. D.**
Professor of Biomedical and
Electrical Engineering
Director of Biomedical Electronics Resource
Case Western Reserve University
Cleveland, Ohio

Based on the Workshop on Solid State Physical Sensors for Biomedical Applications sponsored by the Biomedical Electronics Resource, Engineering Design Center, Case Western Reserve University and the Biotechnology Resources Branch, National Institutes of Health. The workshop was held at Huron, Ohio on December 8-9, 1977.

**CRC Press**
Taylor & Francis Group
Boca Raton  London  New York

CRC Press is an imprint of the
Taylor & Francis Group, an **informa** business

First published 1980 by CRC Press
Taylor & Francis Group
6000 Broken Sound Parkway NW, Suite 300
Boca Raton, FL 33487-2742

Reissued 2018 by CRC Press

© 1980 by CRC Press, Inc.
CRC Press is an imprint of Taylor & Francis Group, an Informa business

No claim to original U.S. Government works

**Library of Congress Cataloging in Publication Data**

Main entry under title:

Physical sensors for biomedical applications.

Based on the Workshop on Solid State Physical Sensors for Biomedical Applications held at Case Western Reserve University, Cleveland, Ohio, Dec. 8-9, 1977.
Bibliography: p.
Includes index.
1. Transducers, Biomedical--Congresses.
2. Medical electronics--Congresses. 3. Human physiology--Measurement--Congresses.
I. Cheung, Peter W. II. Workshop on Solid State Physical Sensors for Biomedical Applications, Case Western Reserve University, 1977.
R857.T7P49     610'.28     79-20056
ISBN 0-8493-5975-9

A Library of Congress record exists under LC control number: 79020056

Publisher's Note
The publisher has gone to great lengths to ensure the quality of this reprint but points out that some imperfections in the original copies may be apparent.

Disclaimer
The publisher has made every effort to trace copyright holders and welcomes correspondence from those they have been unable to contact.

ISBN 13: 978-1-315-89653-3 (hbk)
ISBN 13: 978-1-351-07563-3 (ebk)

Visit the Taylor & Francis Web site at http://www.taylorandfrancis.com and the
CRC Press Web site at http://www.crcpress.com

# PREFACE

The material in this book is based upon a two day workshop on Solid State Physical Sensors for Biomedical Applications held in Huron, Ohio, December 8—9, 1977. The individual sections of the book are based upon presentations made by the authors at the workshop. Each presentation was transcribed and given to the authors for revision. We have also included transcriptions of some of the discussion following each presentation.

This workshop was the second in a series sponsored or cosponsored by the Biomedical Electronics Resource at the Engineering Design Center of Case Western Reserve University. It was supported by the Biotechnology Resources Branch of the National Institutes of Health Division of Research Resources. The editors would like to express their thanks to this agency for its support and especially to Dr. William R. Baker from the Biotechnology Resources Branch for his many suggestions and assistance in organizing the program.

The workshop and this book were made possible by the hard work of many individuals. In addition to the various authors, we would like to thank Mr. Sidney Williams and his staff at Case Western Reserve University for organizing and operating the physical arrangements for the program. We are especially appreciative of their efforts to cope with the problems caused by severe winter weather during the workshop. We also wish to thank Mrs. Cathleen Poon, Lois Schweitzer, and Jean McNulty, for their dedicated work in handling the registration of participants at the workshop and in preparing the transcriptions of the sessions. Finally, we would like to thank the graduate students of the Engineering Design Center who were on hand at the meeting to assist in ways too numerous to mention. All of these individuals played a key role in the success of the workshop, and we dedicate this volume to them.

Michael R. Neuman
David G. Fleming
Wen H. Ko
Peter W. Cheung

January 1979.

# THE EDITORS

**Michael R. Neuman, Ph.D., M.D.,** is Associate Professor of Biomedical Engineering in Reproductive Biology and holds a joint appointment in the Department of Biomedical Engineering at Case Western Reserve University, Cleveland, Ohio. He is also a member of the Engineering Design Center at Case Western Reserve University.

Dr. Neuman received his B.S., M.S., and Ph.D. degrees in electrical engineering from Case Institute of Technology in 1961, 1963, and 1966, respectively. He received an M.D. degree from the Case Western Reserve University School of Medicine in 1974.

Dr. Neuman is a research associate in the Departments of Obstetrics and Gynecology at University Hospitals of Cleveland and Cleveland Metropolitan General Hospital. He is a member of numerous professional societies, including the Society for Gynecologic Investigation and the Association for the Advancement of Medical Instrumentation.

Dr. Neuman's research interests have been in the area of biomedical instrumentation as applied to the reproductive systems in animal models and man. He has published in the areas of biomedical engineering and obstetrics and gynecology, and holds one patent.

**David G. Fleming, M.D., Ph.D.,** is Professor of Biomedical Engineering in Pediatrics and Biomedical Engineering at Case Western Reserve University, Cleveland, Ohio. He is also the Associate Director of the Biomedical Electronics Resource, with responsibility for educational programs.

Dr. Fleming received his A.B. in zoology in 1948 and his Ph.D. in physiology in 1952, both from the University of California at Berkeley. He received his M.D. from Case Western Reserve University in 1973; he was a Pediatric Resident through 1975 and a NIH Fellow in Pediatric Pulmonary Medicine from 1976 to 1978.

From 1968 to 1970, Dr. Fleming was President of the IEEE Group on Engineering in Medicine and Biology. During the same time, Dr. Fleming also served as Chairman of the ASEE's Bio-and-Medical Engineering Committee. He also was a member of the Joint Committee on Engineering in Medicine and Biology from 1965 to 1970, acting as chairman in 1967.

Dr. Fleming's major research interests are biomedical instrumentation for high risk infants and pediatric pulmonary medicine. He has authored many articles in these areas and holds several patents in biomedical instrumentation.

**Peter W. Cheung, Ph.D.,** is Associate Professor of Biomedical Engineering at Case Western Reserve University, Cleveland, Ohio. He is also Director of the Chemical Transducer Laboratory in the Engineering Design Center at Case Western Reserve University.

Dr. Cheung graduated from Oregon State University, Corvallis, with a B.S. in Chemistry in 1968. He received his M.S. in Chemistry in 1969 from the University of Puget Sound in Tacoma, Washington. From 1969 to 1973 he was a graduate research associate with the Center for Bioengineering at the University of Washington, Seattle. He also received his Ph.D. in Electrical Engineering from the University of Washington in 1973.

Dr. Cheung's research interests are in the areas of biomedical transducer and instrumentation for physiological research and continuous patient monitoring. He has published over 30 articles in the areas of fiberoptic oximetry, noninvasive skin reflectance oximetry, oxygen electrodes, chemical sensors, and biomedical instrumentation systems. One of his current major research interests is in the development of chemically sensitive semiconductor devices for biomedical applications.

Dr. Cheung is a member of many professional societies, including the IEEE Group on Quantum Electronics and Application. IEEE Engineering in Medicine and Biology Society, Sigma Xi, the Biomedical Engineering Society, and the International Society on Oxygen Transport to Tissues.

**Wen H. Ko, Ph.D.,** is Professor of Electrical Engineering, Department of Electrical Engineering and Applied Physics, and Professor of Biomedical Engineering. Dr. Ko is Director of the Engineering Design Center, the Microelectronics Laboratory for Biomedical Sciences and the Biomedical Electronics Resource at Case Western Reserve University.

Dr. Ko graduated from the Chinese National University of Amoy in 1946 with a B.S. degree in electrical engineering. He received his M.S. and Ph.D. degrees in electrical engineering in 1956 and 1959, respectively, from Case Institute of Technology, Cleveland, Ohio.

Dr. Ko is a Fellow of IEEE and a member of Sigma Xi and Eta Kappa Nu.

Dr. Ko holds several patents in the area of solid state electronics. His major research interest involves biomedical instrumentation, as well as solid state devices, control systems, and electronic circuit design.

# CONTRIBUTORS

**James B. Angell**
Professor of Electrical Engineering
Director, Resource for Silicon
  Biomedical Transducers
Stanford University
Palo Alto, California

**Douglas A. Christensen**
Associate Professor, Departments of
  Bioengineering and Electrical Engineering
University of Utah
Salt Lake City, Utah

**Jacques Duysens**
Research Associate
Laboratorium voor Neuroen
Psychofysiologie
Campus Gasthuisberg
Leuven, Belgium

**Isaac Greber**
Professor of Engineering
Mechanical and Aerospace Engineering
  Department
Case Western Reserve University
Cleveland, Ohio

**Wen H. Ko**
Director, Engineering Design Center
Case Western Reserve University
Cleveland, Ohio

**Gerald E. Loeb**
Medical Officer, Laboratory of Neural
  Control
National Institute of Neurological and
  Communicative Disorders and Stroke
National Institutes of Health
Bethesda, Maryland

**Ernest P. McCutcheon**
Associate Professor, Department of
  Physiology
School of Medicine
University of South Carolina
Columbia, South Carolina

**Holde H. Muller**
Department of Surgery
School of Medicine
Stanford University
Palo Alto, California

**Thomas S. Nelsen**
Professor, Department of General
  Surgery
School of Medicine
Stanford University
Palo Alto, California

**Robert M. Nerem**
Professor and Chairman
Mechanical Engineering Department
University of Houston
Houston, Texas

**Timothy A. Nunn**
Kulite Semiconductor Products, Inc.
Ridgefield, New Jersey

**James B. Reswick**
Director, Rehabitation Engineering
  Center
Rancho Los Amigos Hospital
Director of Research
Department of Orthopedics
University of Southern California
Los Angeles, California

**Toyoichi Tanaka**
Associate Professor of Physics
Physics Department and Center for
  Materials Science and Engineering
Massachusetts Institute of Technology
Cambridge, Massachusetts

**Bruce Walmsley**
Research Fellow, Department of
  Physiology
Monash University
Clayton, Victoria
Australia

# TABLE OF CONTENTS

*Basic Physics of Physical Sensors*

Chapter 1

# PHYSICAL MEASUREMENTS ON THE HUMAN BODY

James B. Reswick

## TABLE OF CONTENTS

# I. INTRODUCTION

Physical measurements on the human body are important in both diagnosis and research. Indeed, the basic physical examination of the body itself involves the observation and, in some cases, quantitative measurement of mechanical physiologic variables. As more and more technology is applied to medical measurements, many of these variables can be studied in a quantitative fashion employing specific instrumentation systems. Not only can the basic variables of physical diagnosis be studied in this way, but important additional variables can be evaluated quantitatively on an acute or chronic basis. This has allowed and will continue to allow innovative research to be performed in many areas of the life sciences and clinical medicine.

In this paper, I would like to present a comprehensive review of the many different types of physical measurements that can be made on the human body. Rather than discuss a few detailed transducers, I would like to present an overview of the different types of measurements that can be made, and, where appropriate, inject my own comments on problems that can lead to uncertainties in these measurements. The remaining sections of this book will look at many of these areas in much greater detail and present some of the latest innovations in these physical measurements. I would like to consider first the human body as a whole, then the extremities, and finally measurements on internal parts of the body. In view of the great breadth of this subject, I have chosen to present the material in a tabular form. In this way, I hope it will serve as a quick reference for determining the various ways that physical physiologic variables can be measured, and that it will serve as a guide for the life scientist, physician, and instrumentation engineer in determining approaches which might be feasible for solving a particular measurement problem.

# II. MECHANICAL MEASUREMENTS ON THE HUMAN BODY

## A. Gross Body Static Measurements

| Determination | Instrument or method | Comments |
|---|---|---|
| Weight | Scales | Good to a few grams; body weight always changing, evaporation and insensible water loss, etc.; respiration and cardiac action affects most sensitive scales (ballistic cardiogram) |
| Height | "L" bar | Good to a few millimeters; height continuously changes due to body position and tissue loading (discs) |
| Surface area | Estimate from geometric model | Very difficult to measure accurately; effective areas for heat transfer measurements can be estimated |
|  | Mono, or stereo, photo-grammetry[a] | Expensive and time consuming |
| Volume | Water and gas displacement measures | Easy to get moderately accurate measure; lung volume (which varies) not so easy to account for |
| Density | Weight divided by volume | An average, only as good as W and V measures |
| Center of gravity (mass) | Moment equations for two-point support, horizontal table or force plate (vertical position) | Reasonable in supine position and standing on force plate, but obviously c.g. depends on body configuration |

| | | |
|---|---|---|
| Moment of inertia | Estimate from distribution of mass throughout body | Depends on body configuration |

* Stereo-photogrammetry in principle can provide a computer with a complete shape description of the body and/or its parts. Programs to solve for many above quantities (volume, area, centers of gravity, etc.) can be written, and graphic displays of the body from all positions and for cross sections are possible. The technique is not cost effective over other methods at the present time.

## B. Gross Body Dynamic Measurements

| Determination | Instrument or method | Comments |
|---|---|---|
| Center of gravity position in fixed space (position, velocity, acceleration) | Continuous light-spot tracking; TV and Selspot © computer systems | Need c.g. "marker" on body; two or three cameras required for 3-D; continuous parallax correction; fixed reference frame; accuracy inversely proportional to size of field (distance) |
| | Accelerometers | Need stable platform or continuous angle correction for gravity field; hard to fix to body (soft tissue movement); not so hard to fix to top of head to obtain vertical accelerations (cosine relationship helps in tilt error); progressional and lateral accelerations are much more difficult to measure |
| | Interrupted light beams, spaced photo-switches | Gives useful estimate of average velocity, but has no "exact" meaning since body shape changes and velocity varies |
| | Traveling string, rotating potentiometer and/or tachometer | Can produce accurate progressional displacement/velocity record; somewhat awkward for many patients |
| | Footswitches, worn in shoes or attached to bare feet | Indicates events associated with foot-floor contact-velocity, stance, and swing times |
| | Force plates | Give magnitude and direction of foot forces and moments; accelerations of c.g. can be deduced, but c.g. varies as body configuration changes |
| | Stroboscopic still photography | Can give time-position record of body; tedious to analyze; parallax correction may be needed |
| | Still photography of moving light source on body, with pulsed light | Same comment as above |
| | Cinematography and video tape | Gives single frame records of body position vs. time; very tedious to analyze and somewhat expensive in materials |
| Moment of inertia | Accelerate body about c.g. or other known axis | Requires measurement of force (or torque) and acceleration |
| | Swing body as a physical pendulum | Frequency of oscillating known center of mass permits moment of inertia calculation; configuration must be described |

## III. BODY PARTS

Most measurements can be made on the living body (but with associated problems), on cadaver parts, and/or on castings made from plaster or other molds taken from the body. Problems common to most measurements may be itemized as follows:

| | |
|---|---|
| Living body | Soft tissues deform readily; difficult to attach instruments (and know exactly where they are) and to apply forces and moments |
| | Difficult to define boundaries, e.g., where to put plane at elbow between forearm and upper arm |
| | Kinematics of joints are complicated because of changing instant centers and multiple axes of rotation |
| | Landmarks are difficult to define and relate between subjects and over time; soft tissues deform and clothing moves |
| | Configuration of limb affects mass distribution |
| | Mass distribution in 3-D is hard to determine (bone, fats, liquids, muscle, skin, etc.) |
| Cadaver parts | May be difficult to obtain |
| | Properties (e.g., mass distribution) change with time |
| | Difficult to define boundaries; where should cuts be made between parts? |
| Cast models | Soft tissue deformation and muscle contraction alters shape and mass distribution |
| | Models usually are of constant density; difficult to simulate true distribution |
| | Boundaries difficult to standardize |

### A. Body Parts: Static Measurements

| Determination | Instrument or measurements | Comments |
|---|---|---|
| Weight | Scales | Need to know c.g.; difficult to minimize joint moments; very difficult to relate weight to a given boundary; how to define landmarks? |
| | Published tables of % body weight | Good only for average approximations |
| Length | Linear measures | Hard to define, locate, and maintain landmarks |
| Center of gravity (mass) | Multiple scale readings | Joint moments hard to define or eliminate |
| | Published tables | Good only for average approximation |
| Moment of inertia | Known weight distribution and c.g. location | Calculation only as good as original data |
| | Dynamic test as swinging a limb or measuring torque required to accelerate | Hard to relax muscles in vivo; visco-elastic effects at joint can alter results; good when testing cast model or cadaver part, but boundary identification is a problem |

| | | |
|---|---|---|
| Shape, surface area, cross sections | Perimeter measures (circumferential) | |
| | 3-D stereometric light scanning | Stereo-photogrammetry in principle can provide a computer with a complete shape description of the body and/or its parts. Programs to solve for many above quantities (volume, area, centers of gravity, etc.) can be written, and graphic displays of the body from all positions and for cross sections are possible. The technique is not cost effective over other methods at the present time. |
| | Cast molds | (See comment above on cast molds) |
| | Section cadaver parts (frozen or not) | (See comment above on cadaver parts) |
| Volume | Fluid displacement overflow | Hard to detect small differences; hard to define landmarks; difficult if wound is present |
| | Cast model measurements, 3-D stereometric | Easier than with live parts |
| Density | Weight/volume | Both difficult to determine accurately in vivo |
| | Cadaver disection; frozen sections | Enables determination of tissue-bone densities and to create maps |
| Pressure | Pressure distribution on feet when standing, or on buttocks, etc., when sitting; glass plate on rubber cones with camera, pressure indicating plastic, array of pressure transducers (foam, strain gauge, capacitive, pneumatic element with switch, etc.) | Weight is transferred through viscoelastic tissue; therefore, transducer must be small and thin so as to not change contour; a large array gets expensive and difficult to read out, especially dynamically |
| Shear | No good system known to author | |

## B. Body Parts: Dynamic Measurements

| Determination | Instrument or measurements | Comments |
|---|---|---|
| Forces and moments | Torque measures of reflex response, stimulated (electric) muscle characteristics, and voluntary muscle action | Dynamometers give continuous records of torque; can show strength and time delay of reflex following tendon tap |
| | External loading machines, e.g., Cybex® | Difficult to align centers of rotation along instant centers of joints and to apply loads accurately to/through soft tissue; loads may depend on voluntary effort of subject |
| | Controlled velocity and acceleration | "Passive" machines not too difficult; active machines require subject to voluntarily (trained response) control reactions, e.g., constant velocity requires tracking with training |

| | EMG surface and implanted electrodes | Processing of EMG signals still under research; correlation of force and "magnitude" of EMG for isometric contraction is fair (10%), but no correlation has been documented for dynamic (isotonic) or mixed, normally moving) muscle activity; useful to determine if muscle is "on" or "off" |
|---|---|---|
| Kinematic measurements (instant centers, element lengths, joint angles, location of c.g.) | Goniometers | Must be designed to accommodate changing instant centers; should be lightweight, strong, acceptable to subjects, and easy to use |
| | | Soft tissue movement is a problem; reference landmarks are difficult (bony prominences, centerlines of bones, or what?) Readout and recording are straightforward Hip and shoulder joint angles are especially difficult; reference coordinate systems must be defined; hard to attach 3-D goniometer stably to body |
| | Light spot tracking; computer analysis, graphic display, and rapid printout of data and pictures | Landmarks difficult to define and attach (soft tissue and clothing movement) |
| | | System must accommodate parallax in three dimensions; "cross over" ambiguity of multiple tracking points; accuracy is a problem when large field is desired; if camera moves with subject, reference frame is difficult; systems are expensive |
| | Accelerometers | Linear and angular accelerometers can be attached to body parts; need six coordinates to define position of rigid body in space, i.e., six accelerometers; gravity field must be accounted for in linear accelerations |
| Pressure | | See comments under "Pressure" Static Measurements |

| Shape Changes | Muscle bulge detectors; switches, fluid-filled bags, etc. | Not hard to produce signal, but difficult to hold in place over a long time |
| | Stretch detectors; switches, linear transducers (strain gauges, magnetic field detectors, potentiometers, etc.) mounted to harness or to points on skin; mercury-filled elastic tubes; pressure-sensitive foams (as switches and proportional measures) | Practical for shoulder-operated prosthesis control (multilock switches and proportional devices), more difficult when end points must be fastened directly to skin for long periods; circumferential changes can be indicated by mercury-filled tubes, but hysteresis and changing properties make system inaccurate; pressure-sensitive variable resistivity foam makes reasonable switch, but poor on proportionality |

# IV. SKIN AND INTERNAL PARTS (ORGANS)

## A. Skin and Internal Parts (Organs): Static Measurements

| Determination | Instruments or measurements | Comments |
|---|---|---|
| Volume, shape, cross sections, relative locations (bones, muscles, heart, brain, tendons, etc.) | X-ray; plain, dye injections, barium in gut, air in brain, etc. | X-ray justified in diagnosis, but should be limited for research studies; some visualization techniques (e.g., dye injection) carry risk |
| | EMI, (CAT), Xerography, Xonics | |
| | Ultrasound | Cross sections difficult in vivo, although ultrasound and EMI give significant data |
| | Nuclear scans of radioactive take up | |
| | Cadaver dissections, frozen slices, etc. | |
| | Computer graphic visualization (3-D landmarks from X-rays) | Tedious to measure and enter data; good 3-D visualization of result |
| Physical parameters (including dynamic measurements); stress-strain, viscoelastic, strain rate, strain energy, mechanical impedance properties | Specimen tests in vitro | Samples must be as "vivo" as possible, quick frozen, controlled environment |
| | Tissues, bones, etc. of animals may be exposed to transducers | |
| | Skin may be tested using contact transducers, linear and torsional; dynamic impedance tests and ultrasonic impedance transducers | Hard to develop base lines; i.e., what is "0" starting point?, how to reproduce?; edge, end, and underlying tissue effects hard to control |

| | Instruments or measurements | Comments |
|---|---|---|
| | | Skin variation (age, type, etc.) hard to standardize |
| | Bone; properties, shape and distribution of structural elements (cortical vs. cancellous); slow (quasi-static), fast (strain rate controlled) and impact tests; low and high frequency (ultrasonic) measures of fracture healing | Hard to test in vivo; vitro tests require carefully controlled preparations and environment Coupling through soft tissues is hard; both input and output |

## B. Skin and Internal Parts (Organs): Dynamic Measurements

| Determination | Instruments or measurements | Comments |
|---|---|---|
| Position, velocity and acceleration, translation and rotation; bones, tendons, muscles (organs: heart, diaphragm, gut, etc.) | Cinema radiography without and with markers (wire, radiopaque dye) | Gives qualitative information, but tedious to analyze frame by frame |
| | Serial X-rays Scanning ultrasonics | Reference planes hard to maintain |
| Forces and moments | Bones and joints; deflection of bone (X-ray), force (pressure) transducers in joints (head of femur), strain gauge on bones or on implants (Harrington rods, plates, hip prosthesis), telemetry or hard wire | Invasive techniques have risk; telemetry is tricky, but has produced results; hard wires good for short time only |
| | Tendons; direct measure of muscle force; strain gauge on tendon or on series strain element during surgery; transverse pull on tendon with angle measurement | Invasive, therefore risky; very hard to measure tendon force in vivo, especially over extended periods |
| | Deduce from external force and moments | Poor accuracy due to tissues, joint friction and other structures; difficult to determine moment arms accurately |
| | Muscle; mainly from tendon forces and deduction from external loads; rough correlation between area cross section and max-force exists | See "tendon" comments above that apply to muscles as well; pinnate muscles have herringbone fiber structure complicating determination of active cross section |
| | EMG as a measure of dynamic force | See comments under "Body Parts Dynamic Measurements"; very poor quantitative measure of force when muscle is moving |
| | Joint friction or hydrodynamic effects; sinusoidally rotate limb, plot torque vs. position; Lissajous figure shows hysteresis, therefore, friction loss | Useful for finger joint studies; will work on other joints; need good machine to measure quantities and accommodate changing instant centers |

| | | |
|---|---|---|
| Pressure | Fluid compartments; circulatory system;[a] sphygmomanometer, external pressure transducers, in-dwelling pressure transducers via catheters | External systems good for moderate accuracy, but tissues always cause distortion of signal; implant systems fine in large vessels, but are invasive; very small vessels present problems; blood is a nonNewtonian fluid, especially in small vessels |
| | Lymphatic system[a] | Hard-to-locate pools, pressures are very low |
| | Muscles and other fibrous structures; manometers (static or low flow pump systems with transducers) | Chambers are not pure fluid pools; structures clog openings; most data show wide inconsistencies |
| | Spine and brain; taps with manometers or pressure transducers | Good readings, but invasive |
| | Disc; taps with low-flow pump and transducers | Probably meaningful in young, healthy disc; older and degenerative discs are fibrous throughout, readings uncertain |
| | Eye (glaucoma) tonometers | Practical |
| | Urinary system; catheters with pressure-sensitive bulbs | Useful for diagnosis |
| Flow | Blood; implant flow meters, ultrasonic doppler, tracers and dyes, magnetic induction, thermal dilution (cold fluid migration) | Invasive and implant systems pretty good; noninvasive systems are less reliable; very hard to get velocity profile; arterial flow easier than venous; total heart output is important, but hard to measure accurately |
| | Lymph; time of tracer outflow from point of injection to lymph node, decay rate at point of injection, and build-up rate at lymph node | Hard to get repeatable results |
| Shock waves | Cranium; calculation from fluid properties, bone properties, and configuration are possibly meaningful; damage tests on primates, or models give some data, accelerometer on bone gives measure of shock following impact and timing | |

[a]   In all of the above, micro pressure transducers with electrical outputs should be considered. Output by hard wire or telemetry.

Chapter 2

# DISTURBANCE PHENOMENA IN PRESSURE MEASUREMENT

## Isaac Greber

## TABLE OF CONTENTS

# I. INTRODUCTION

The editors gave me the opportunity to edit my talk for this publication. After some soul searching, I resisted the temptation to rewrite the talk to make it sound like a paper and to let it reflect what I've learned since presenting the talk. Instead, I am leaving it as it was, a conversation I had with the participants at the meeting. I've removed a few of the conversational warts, the ones that the ear accepts, but the eye rejects. However, despite vanity I have left most of them in. I hope this helps preserve the atmosphere of a meeting in which we learned from each other in imperfect, but stimulating, conversation.

# II. PRESENTATION

I decided rather early on in thinking about this talk that it would be rather presumptuous of me to speak about the fundamentals of pressure measurement with this audience. Instead, I decided to discuss one aspect of the interaction of pressure measuring devices with the things to be measured, which has been of concern to me. Although what I will talk about will be in some sense a fairly elementary question and quite specific, I think it is useful to consider it on a somewhat broader scale: the obvious idea that any device that we use to make a measurement interferes with the things to be measured so that we do not measure what we intend to measure. Thus, it may be important to consider this interaction in interpreting the measurements.

Figure 1 is a very simple example of what I am referring to as a weak and a strong interaction of a measuring device with the quantity to be measured. I'll consider that we want to measure the pressure $p_o$ within some chamber whose volume is $V_o$, and I want to imagine a transducer located at the side of that chamber. Let's make believe that perhaps it looks a little bit different from this simplified sketch in that there is a valve between the chamber and the transducer such that we have a very tiny volume between the two. We calibrate our transducer, equalize the pressures across it, and then expose it to the internal pressure. The transducer deflects some volume $\Delta V$. Suppose I take the very simple idea that the process of change of volume occupied by this gas as the transducer deflects it as an isothermal process so that the product of pressure and volume of the gas is the same before and after. Then, if we think of a weak interaction, I first write the correct expression stating that the pressure in the initial container is equal to the pressure that is measured by the transducer multiplied by 1 plus the volume change divided by the initial volume (Equation 1).

$$p_o = p_i (1 + \Delta V/V_o) \tag{1}$$

We neglect the volume change by making a sufficiently stiff transducer so that the new pressure is almost the original pressure. If we have a strong interaction, it means we haven't made a transducer stiff enough. We then continue to use the correct expression relating the pressure in the chamber to the pressure that is measured by the transducer.

The important thing to keep in mind is that the transducer operation really provides us with two pieces of information, a calibrated piece of information that relates the deflection change to a pressure difference across the transducer, and an uncalibrated piece of information that tells us how much the transducer itself has deflected. We normally use just one piece of information, we use the calibrated part of the information, but we have available to us the other piece of information, the actual deflection of the transducer. Suppose we make that measurement twice, use two different transducers. I'll call these two different transducers 1 and 2, and now, using two measure-

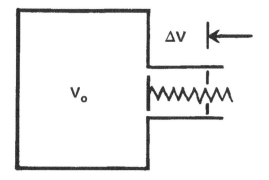

**Weak:**   Po = P(1 + ΔV/Vo) ≈ P

**Strong:**  Po = Pi(1 + ΔVᵢ/Vₒ)

$$P_o = \frac{P_1 P_2 [\Delta V_1 - \Delta V_2]}{P_1 \Delta V_1 - P_2 \Delta V_2}$$

FIGURE 1.   Pressure measurement with weak and strong interaction.

ments, relate the pressure in the chamber before the deflection to the two pressure measurements given by the two transducers and the two volume changes in the course of those measurements. The relation is given in Equation 2:

$$p_o = \frac{p_1 p_2 [\Delta V_1 - \Delta V_2]}{p_1 \Delta V_1 - p_2 \Delta V_2} \tag{2}$$

What we are doing is saying that even if we have a transducer that deflects a relatively large amount in comparison with the initial volume, we can still use it, provided that we make more than one measurement. So, the strong interaction is useful, and one typical way of using a strong interaction is to account for the process. Here I'm saying, "Let's make believe the process is isothermal and make additional measurements that allow us to use the information from the strong interaction to deduce the initial pressure."

Now there is a catch to all of this, other than the idea that I had to identify the process that occurs. Suppose we want to do something with the pressure level that occurs in this chamber. Then we better make sure that after we finish the two measurements we return the chamber to its initial pressure. This means, more generally, that if we're going to make a measurement using a strong interaction, the interaction better not destroy the process that we are interested in in the course of making the measurements.

I hope this sets the ideas. I now want to consider one example which has some of the characteristics that are common in many of our pressure measuring techniques. That is, a pressure transducer in interacting with biological surfaces typically meets a surface that is deformable. There frequently is a liquid layer between the transducer surface and the surface of interest, and frequently, it is necessary to place the transducer's sensitive element within a small cavity. A simple example that I will show will have at least these three features in it.

For the purpose of focusing on this, I'm considering a nonbiological situation (Figure 2) that may be a rough approximation to some biological ones. We have a balloon

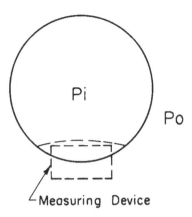

FIGURE 2.  Pressure  measurement
on free balloon.

with some pressure inside, some pressure outside, and we have a measuring device with which we push a little bit into the surface of our balloon. One of the situations in which you use such a measuring device was referred to by Dr. Reswick, and that is in tonometry. Here, the problem is that you would like to measure the pressure inside the eye, actually it's in the aqueous humor, by taking a device that pushes against the eye, makes the surface locally plane, and then, because that surface is plane, saying that we have taken up the load by the transducer. I want to distinguish in this situation between two kinds of use of the transducer, a dry contact and a wet contact. If you have a dry contact and you push in, think of the balloon or model eyeball. The thing we imagine is the surface flat against the flat surface of the transducer. If we merely have a tensed surface, and there are no other structural properties that concern us, then with a flat, tensed surface resting on the transducer, the transducer will be loaded uniformly across its entire surface. If we think of a wet interaction, that is, think of tear fluid between the transducer and the surface, then the tear fluid is squeezed out from between the two surfaces, and a pressure distribution is set up. The shape of the deformable surface is not a flat line, but some curve, and the curve is determined such that the right pressure distribution occurs. I've indicated two lines that are intended to point out that inflection points exist, and that those inflection points may be radially as far out as the edge of the transducer, or may be radially inboard, because the shape is three dimensional and the principal radii of curvature may be of opposite sign. The detail surface shape and the pressure distribution are related. The important qualitative feature is that the transducer is not uniformly loaded.

What I have plotted in Figure 3 is the pressure distribution on the transducer surface as a function of the shape that occurs between the transducer surface and the deformable surface of interest, recognizing the fact that the liquid is flowing radially during this process and that a pressure distribution, therefore, occurs. If the two surfaces are flat, as shown in the middle sketch on Figure 3, then the distribution of pressures will be parabolic, as indicated by the solid line in the figure. If the curvature is such that a smaller gap occurs in the middle than towards the periphery, then the pressure will be high in the middle and decrease to its value at the periphery. If the gap is larger in the middle, then the pressure distribution will be flatter. I've indicated in the figure what the average is. For the parabola, of course, the pressure in the middle has a value that is twice the average, and the effect of shape change is radical. Therefore, the pressure that might be measured by a piece of surface that doesn't go the full distance can be quite different from what we think the pressure in that eyeball is. The important thing

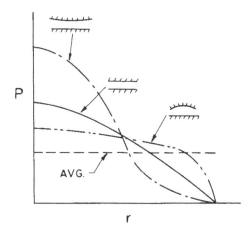

FIGURE 3. Pressure distribution in liquid films as
a function of gap shape.

to carry away from this sketch is the idea that the pressure distribution depends radically on shape, and fundamentally, the shape is not something that we know for a
deformable membrane. The pressure distributions for nonparallel surfaces depend
upon the distance between the two surfaces, whereas the pressure distribution for parallel surfaces is always a parabola. The consequence of this with deformable membranes is that the pressure distribution not only is not known because we do not know
the shape, but also, for any given loading, it will change with time as the size of the
gap changes with time. Therefore, if you push onto an eyeball, the pressure distribution will depend upon not only the pressure inside, but also the force as it varies in
time. If you considered just the difference between the flat surface and the average,
you have a difference of a factor of two. If you start making convex surfaces, then it
is very easy in situations, for example, typical of contact lenses, where I know some
of the numbers, to go up to factors of five or six with no trouble at all.

Fortunately, in lots of applications to pressure and measurement you tend to go
towards the flatter distributions. Therefore you have errors that might be of the order
of magnitude of 50%. That is probably a typical number to keep in mind.

The distribution depends not only upon whether the transducer surface and the surface on which you want to measure the pressure are parallel, but actually on their radii
of curvature. This would mean that for a surface that is much more complicated than
a membrane, and of course we are not talking about membranes, we are talking about
biologically tethered surfaces which have structural properties that are seriously different from pure membranes, we would have situations where even shapes that look parallel might imply more complicated load conditions. Now, if you have two parallel
surfaces that are not flat, then the distribution is not parabolic. It is something else
and depends on the radius of curvature, and therefore, it will also depend upon gaps.
If you have two parallel surfaces and move them with respect to each other, along
with what appears to be a radial line, the pressure distribution will vary in time.

I have indicated in Figure 4 the kind of thing that happens if we think of how a
surface deforms when it is near a solid surface. So, I'm thinking we start off with a
flat transistor and a flat membrane, and ignore the fact that we really should talk
about membranes that have structural properties other than tension. If we have two
parallel flat surfaces, then the fluid mechanics of the situation says that the pressure
distribution is parabolic. If the membrane is strictly a tensed membrane, then the pressure difference across it must be constant, and if the pressure above the membranes

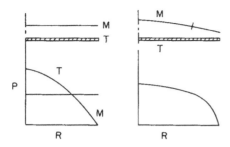

FIGURE 4. Interaction of transducer, T, with deformable membrane, M.

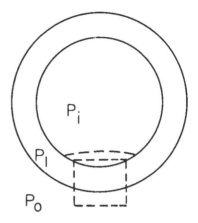

FIGURE 5. Pressure measurement on enclosed balloon.

were constant, the implication, if we looked at the membrane itself, is that the pressure in the gap is constant. The membrane then takes on a shape such that the pressure distribution determined by the fluid mechanics and the pressure distribution determined by the pressure difference across the membrane match. And that's what I tried to indicate in Figure 4.

Some of the pressure distributions can be calculated for simple structural properties. Even in the case of two dimensions with a membrane whose structural property is simply that it is tensed, the matching involves a solution of a nonlinear integral equation for the shape. It can be solved in a fairly straightforward way by successive approximations. If the structural properties of the membrane are more complicated than simply being tested, then correspondingly, we have a more complicated problem to be solved, but at least we can set it up. This gives us an idea of what is happening in that interaction process. I have also indicated, once again, that typically what we expect is that an inflection point exists on the deformable surface, contrary to what one would expect in two dimensions.

Now in many of the measurements that we are dealing with, the situation becomes more complicated in the sense that we don't just have an exposed surface, such as the exterior surface of the eye. One of the things that I am very much interested in is measurements of pressure in the pleural space surrounding the lungs. The picture as shown in Figure 5 is again a balloonlike picture to set ideas. This balloon is encased in another surface with a fluid that has a different pressure from that in the interior of the balloon and from that of the outside. What we would like to do is insert a transducer into the gap and measure the pressure levels in it. We find that when we put a transducer in, it deforms the interior surface.

I've tried in Figure 6 to indicate the problem in a little closer perspective. The upper line that says "to scale", points out the fact that frequently we are talking about surfaces that are so small that if we place a transducer in, anything that we are going to do is going to make a big disturbance. There is no way of placing a transducer in many of our biological gaps without making a radical deformation. I have exaggerated the scale to point out what we do. The transducer radically distorts the surface, and you will note that I can't show all of the transducer in the figure since it is off scale. As a practical matter, you wind up not being able to make the sensitive surface occupy the whole region, so you have a sensitive surface which is someplace in the middle. You have a large surface, larger than the transducer pressure-sensitive region, and you have a liquid layer between the surface that you are interested in and the pressure trans-

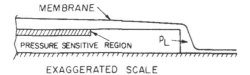

EXAGGERATED SCALE

FIGURE 6. Large deflection caused by "small" transducer.

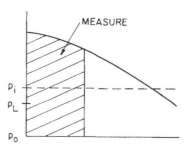

FIGURE 7. Force on sensitive region in nonuniform pressure distribution.

ducer. What you would really like to do is measure the pressure in this liquid when the transducer wasn't there. What you wind up doing is measuring something. To get an idea of what it is that you measure, let's draw a pressure radius diagram again, and let's claim that we already know what the pressure distribution is. We don't, obviously, because, again, that is a distribution question determined by the interaction of the transducer with the deformable surface. We have unknown pressures, $P_L$, (Figure 7) at the boundary of the transducer device, we have some pressure, $P_i$, that is supposed to denote an average, and we have an unknown pressure distribution. We measure some grand total force on our sensitive region, and from this deduce something that is supposed to be a pressure that appears in the absence of the transducer. Well, if we're really lucky, then the grand total force on the transducer may be something close to what it was before we put it in the gap. This requires that, although the transducer clearly makes a gross disturbance of the situation locally, it is only local, and the grand total force is all right. If that is true, if we were able to insert a large transducer, we'd measure what we're interested in. So, if we were able to satisfy all my ifs, and you can't, there is still one more if. If we knew the distribution, we'd be able from our measurement to infer what happened in the absence of the transducer. Well, at minimum this gets back to Figure 1. You need more than one measurement. You need enough measurements to define the distribution. Notice what happens then. Suppose you had a transducer with a number of measuring sensitive areas on it. Then you could define a distribution, and then, despite the fact that you couldn't make a transducer that occupied the full region, perhaps you could get an idea of what the total force on that transducer device is. Once you recognize that the interaction must occur, you see that it doesn't help to make a tiny device, because a tiny device on the scale of the measurement is a big device. It may make more sense to make the device bigger so that you can make more than one measurement on the device and get an idea of distributions.

I have focused on one elementary problem, and it's probably something that we all know, but our thinking about it is not usually in the forefront. I believe the implications are much more serious. In this kind of transducer application, in lots of other pressure measurements, and in many other measurements, we find in biological systems that there is absolutely no way of getting in there and making a small disturbance. The thing to do may be to give up trying and make enough big disturbances so that from the combination of measurements you can deduce the quantities that you'd like to measure.

Let me conclude by pointing out a few things I have left out of the simple example. When you have thin gaps, one of the things that happens is that small flows induce very, very large pressure differences. The result, then, is that when you make a large disturbance, which is fundamental to anything that is put into a gap, and try to infer quantities that occur in the absence of that disturbance, you change the flow patterns that exist. So, the things that I said that sound so nice as long as you don't have interference with the flow patterns become radically altered when you recognize that the pressure measuring device, because of its interference with the flow patterns, experiences on its peripheries a pressure, $P_L$, which is unknown. You then have to go to the next level in the examination of what is happening, examine the interaction of the pressure measuring device with the flow patterns that occur in the small gaps in order to get an idea of the distributions, and complete what in my simplified approach of a static external situation appears to be a completed problem.

# DISCUSSION

**Dr. James Reswick:** In the case where there is fluid which causes the distortion in the small space, what maintains the pressure in that fluid? For example, the tear-drop glaucoma test, why don't the tears just wash away when you come up against the surface of the eyeball?

**Dr. Greber:** They do. It takes a long time; and that long time is very, very long in comparison with the time of the tonometry examination. This doesn't cause any problem in tonometry, none whatsoever. The reason is that because the whole thing is exposed you can have a transducer surface that occupies the entire area that you are pushing against the eyeball. Therefore, although the transducer is nonuniformly loaded, you are measuring the grand total force on the transducer. That's fine, that is really what you are looking for, and you then get a correct average and everything is fine. The problem in many of the other biological applications is that you do not have the freedom to use a transducer, at least at that level, that is sensitive over its entire surface.

You alluded to some of these problems when you talked about trying to put transducers against the skin. You fundamentally packaged something. You had that kind of problem. In all of these cases where you try to put something inside, you make a package. So, your interaction region is different from the sensitive region, and the forces that you experience on the sensitive region are quite different from what might be indicated by the averages.

I should point out one other thing, and that is, I have indicated pressures without indicating signs. If I think of these as magnitudes, and if I think, for example, of trying to measure the pleural pressure, I have a pressure difference between the center of a transducer and the outside whose sign changes in time. The curves I sketched remain correct for the magnitudes, but the sign changes. This means that what you would indicate by a pressure measurement might be something like the doubling of the excursion of pressures as you go through a breathing cycle. The same thing would happen in the surface measurements on the skin. You might wind up concluding that a force goes through a cycle that might be something of the order of twice what it really is. You don't get rid of liquid layers even if you use glue. Glue is a highly viscous liquid. The mechanism by which pressure-sensitive tape adheres to a surface is precisely the mechanism that I am talking about. I've talked about simple lubrication ideas as they influence pressure measurement and as they influence the operation of both measuring devices and the biological functions themselves. When you use glue, the glue itself gives you a mechanism, even if you thought you're dealing with dry surfaces, of manufacturing nonuniform loading on pressure transducers. If you want the grand total force, and if the pressure transducer surface occupies the whole surface, fine. If it doesn't, then you're in trouble, and you better account for it. However, accounting for it can be easy if you are willing to make a transducer large enough to get enough information to get an idea of distribution.

**Dr. Lee Baker:** I gather that your problem is in measuring intrapleural pressure, that is the pressure between the lung and the chest wall.

**Dr. Greber:** Well, that's one of the things that I am concerned about. I am also working on eyeball problems. I think that the problem, however, is rather pervasive. It occurs if one is interested in tissue pressures. It occurs if you want to measure lymphatic pressures. It is all over the body, there are lots of situations.

**Dr. Lee Baker:** I'm just wondering in regard to intrapleural pressure, if you have any data that might be compared to, say, the values we normally quote, and if so, what they are; and if not, could you speculate on what the distribution of pressure might be? How could we make this measurement accurately?

**Dr. Greber:** That one is easy for me to discuss in terms of distribution, and less easy in terms of pressures. One of the things I am working hard to understand now is exactly what people are measuring when they measure intrapleural pressures. They do things just as I sketched. They are measuring things that characterize a loading, but what to infer from these findings is not clear. They may be measuring precisely what I said, excursions that are twice the order of magnitude of the real excursions. Now we get to the question of distribution. Measurements indicate that distributions of intrapleural pressure are nothing like hydrostatic. The reasons are, again, squashed layers, slow flows, large pressure differences associated with low flows.

I prefer to tell you orders of magnitude only from calculations that I've done because of my problems of making inferences from the measurements. You may have pressures during the breathing cycle that may be close to being constant over the lungs, whereas your first bad guess might be that they are close to hydrostatic. The combination of slow flows and deformable bodies, I think, in some ways can be thought of as being a nice equalization process to make distributions a lot more constant than you might otherwise guess.

**Dr. Lee Baker:** Do you think that due to gravity forces, for example, that one does have the distribution of intrapleural pressure, say, from higher at the bottom of the lung to less at the top? This is used to explain regional ventilation.

**Dr. Greber:** That's really one of my major points in some of the work I am doing. That is not true. And the reason is precisely the slow flows that will occur between the lungs and the chest wall. You can get, with very tiny flows, almost complete cancellation of the hydrostatic pressure distribution.

Chapter 3

# PHYSIOLOGIC FLOWS AND THEIR MEASUREMENT*

Robert M. Nerem

## TABLE OF CONTENTS

# ABSTRACT

The measurement of physiologic flows is of interest both from the viewpoint of clinical diagnostics and in order to gain a better understanding of the various organ systems for both physiologic and pathologic conditions. Such measurements may range from that of quantifying the general flow rate in a system to documenting the nature of very specific, local, detailed, fluid dynamic events. In either case, the extreme complexities of such flows are a limiting factor on the measurements which one can perform. Physiologic flows are, in general, not simple. The flow complexities that can arise include asymmetric velocity profiles or flow patterns due to branching and curvature effects, turbulence, secondary flows which manifest themselves as a corkscrew, helical motion associated with streamline curvature, flow separation in which the downstream moving fluid separates from the wall and subsequently reattaches resulting in a dead water region, and the pulsatile or time-varying nature of a flow. In this presentation, these flow features will be discussed using, primarily, the cardiovascular system as an example. The measurement of physiologic flows is also limited by instrumentation characteristics, and the interaction between the nature of physiologic flows and instrumentation currently in use also will be discussed.

# I. INTRODUCTION

When I first was invited to participate in this meeting, I was asked to present some general introductory and tutorial comments about physiological flows and their measurement. Although in some sense it probably was expected that I would provide a state-of-the-art survey on how one measures flows in all the various types of physiological applications in which one might be interested, I have opted not to do that.

Instead, I have decided to provide a somewhat different perspective, and I will start by stating what I consider to be a very basic fact of life: physiological flows are extremely complicated. Such flows are very complex in nature, and as a result, our ability to measure such flows is limited. This limitation is imposed not only by the frequency response of the instrumentation system one chooses, but also by the ability of that system, whatever it is, to discern the actual events that are taking place and the ability of the user to discern what the output of the instrument is trying to communicate.

In the performance of any measurement, there is a requirement that the instrumentation system have a response, both static and dynamic, that will allow one to follow the events taking place, and also, that there be minimal interference with the phenomena being observed. However, it is also necessary that one understands, e.g., in the case of flow measurements, the kind of fluid dynamic events that are taking place. In general, what we don't know about and what we don't recognize as a possibility, we do not see. The result is that very sophisticated instrumentation systems can be and have been used to produce very simple and, in some cases, almost simple-minded observations. Because our ability to observe a flow experimentally is not limited by pure instrumentation characteristics, but also by the complexities of what we are trying to observe and by our ability and willingness to perceive those complexities, I wish to address these complexities — these intricate facets of fluid dynamics as they manifest themselves in physiological systems — in this introduction.

# II. PRESENTATION

Our knowledge of basic fluid dynamics, as it relates to these kinds of systems, comes from a variety of places. It comes from model studies, it comes from mathematical studies, and to a limited degree, it comes from in vivo studies. With regard to the latter, in vivo verification has primarily been obtained from the cardiovascular system, particularly from blood flow in large arteries. Since that happens to be my own specific

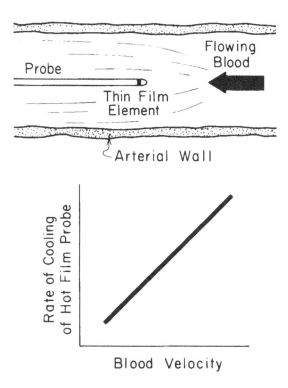

FIGURE 1. Use of hot-film anemometer probe in the measurement of blood velocity. Both the positioning of the probe and the fact that the rate of cooling is dependent on blood velocity are shown.

research interest, the illustrations I have chosen are from that area. However, I hope that each of you will think about some of the details to be mentioned in terms of the kind of fluid system and organ system in which he or she might be interested.

Any extensive in vivo verification of physiological fluid dynamic concepts, as indicated earlier, is limited primarily to the cardiovascular system. In that area there are two instruments whose use has resulted in primary contributions to the relatively recent acquisition of at least some detailed knowledge of large-artery blood-flow phenomena. These are the hot-film/hot-wire anemometer technique[1-3] and the pulsed ultrasonic Doppler technique.[4-7] Both of these instruments provide a measurement of point velocity, i.e., the ability to obtain a time-resolved point velocity measurement. There may be many cases where all one is interested in is time-averaged, spatially integrated information. However, in terms of some of the complicated fluid dynamics present in physiological systems, one ultimately would like time-resolved point velocity measurements, and these two instruments have such a capability.

In Figure 1, a hot-film velocity probe positioned in a blood vessel is illustrated. The probe could be on the end of a catheter or it could be of an L-shape configuration and placed in position with the chest open by coming in through the vessel wall. The probe has a heated thin-film element which is part of an electrical circuit and is maintained at a constant temperature. As blood flows over the probe, it tries to cool the film. The amount of energy that has to be provided to maintain the film at a constant temperature can be monitored with the net result that, since there is a relationship between the rate of cooling and the blood velocity, one can relate that rate of cooling to the electrical power being provided and obtain a calibrated output.[8] Such probes,

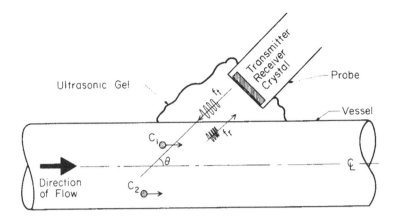

FIGURE 2.   Pulsed ultrasonic Doppler flow meter illustrating transmitted and reflected waves and scattering blood cells ($C_1$ and $C_2$).

whether they have a film such as shown in Figure 1 or a small-diameter wire on the order of a thousandth or ten thousandth of an inch in diameter, can be used with relatively high frequency response characteristics. Hot-films can sense frequencies up to 500 or even 1000 Hz,[9] and hot-wires, which are more normally used in gaseous fluids than liquids, can sense frequencies up to thousands of Hertz.

The pulsed Doppler ultrasonic velocimeter is illustrated in Figure 2. Shown here is a transmitter/receiver crystal sending out a short wave train of ultrasonic waves with a frequency, $f_t$. It is coupled with ultrasonic gel to a vessel. At some fixed time later, the crystal receives a portion of that wave that has been reflected from a specific position and now has a frequency, $f_r$. The time delay involved between pulse generation and when the window is opened to look at the reflected signal provides range information because one knows the propagation velocity in the system. In this manner, one measures a velocity component, and if one makes an assumption about flow direction (the angle $\theta$), then one has the flow velocity at the point corresponding to the range associated with the received reflected signal.

Figure 3 illustrates centerline velocity waveforms measured with these two different types of instruments, where measurements have been performed simultaneously in very close proximity to one another in the abdominal aorta of a horse.[10] One can see that there is rather good agreement in this case. The peak velocity is on the order of 25 cm/sec, and although there are some differences, the waveform shapes compare favorably.

I have briefly discussed these two instruments because they have provided in vivo verification of the kind of detailed fluid dynamics to which I now wish to address myself. Figure 4 is intended to illustrate the classical type of flow about which we hopefully all know. It is a fully developed Poiseuille flow which is laminar and has the characteristic parabolic profile. If one were making velocity measurements in the Alaskan pipeline, one might find that this was a reasonable representation of the flow pattern. However, this is not the case in physiological flow systems. Certainly, none of the ones in which I have been interested have demonstrated the existence of this type of flow. As far as the larger arteries of the circulatory system, Poiseuille flow is not what is present, and although there are many reasons for this, this is primarily the result of the flow not being steady and the pipes not being straight nor very long. Because of the latter, there are entrance effects as illustrated in Figure 4. Instead of Poiseuille flow, there are a variety of complexities to be encountered. These do not make life easy for the experimentalist, but they do make life interesting. Here, I only

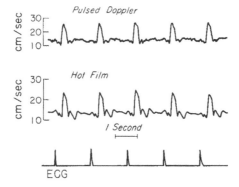

FIGURE 3. Comparison of pulsed ultrasonic Doppler and hot-film anemometer measurements of centerline blood velocity in the abdominal aorta of an anesthetized horse.

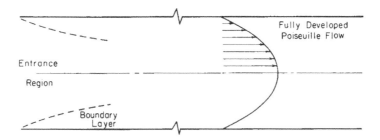

FIGURE 4. Laminar pipe flow, including both the entrance region and the downstream region where fully developed, Poiseuille flow with a parabolic velocity profile is present.

have time to discuss some of the intricacies. Again, try to keep in mind what influence these phenomena, these details, might have on measurements you would like to peform.

One of the first flow characteristics that one observes in just about any measurement of large artery flow (it would also be true of flow in the airways and, probably, in numerable other examples in physiology) is that the velocity profile is asymmetric. As shown in Figure 5, this might be associated with a branching situation or with a region of sharp curvature. However, in general, the geometry almost mandates that there be nonsymmetry in the flow pattern. Furthermore, although it is illustrated in Figure 5 that one has the reciprocal of the profile on the one side of a bifurcation also on the other, that is not necessarily true. It only would be true if the geometry were exactly the same and if the flow were evenly divided. However, downstream peripheral resistance may not be the same on the two sides, and the flow division may be very uneven. Thus, even though the geometry may be the same on both sides, if the flow does not evenly divide, there will be a different pattern in one branch than the other.

Let us consider now two examples of a curved pipe, one being the ascending aorta and the other being the left common coronary artery. In both cases, I will discuss velocities in the region of the inside wall vs. the outside wall in terms of the kind of asymmetry that is present. The data of Seed and Wood[11] for the canine ascending aorta velocity profile are presented in Figure 6, where measurements of velocity obtained at peak systole are correlated as a function of position across the lumen of the

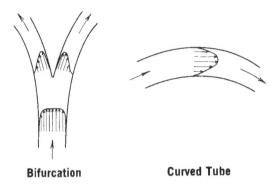

**Bifurcation**        **Curved Tube**

FIGURE 5.   Illustration of the type of asymmetric velocity patterns which may be present in a region of bifurcation or where there is sharp vessel curvature.

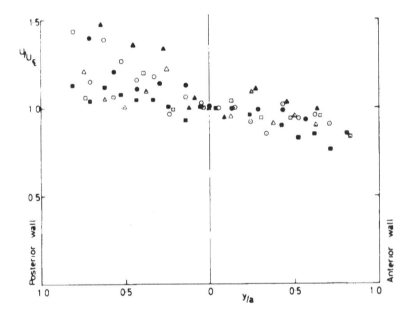

FIGURE 6.   Nondimensional velocity profile in the ascending aorta at peak systole
as obtained by Seed and Wood[11] in anesthetized dogs. Measurements were performed with a hot-film anemometer in the plane of curvature of the aorta.

vessel. From Figure 6 and noting the inside or posterior wall of the arch and the outside, anterior wall, one can see that at peak systole, and within the scatter of the data,
there is a definite change in velocity as one moves from the inside wall to the outside
wall, with the higher velocity being at the inside wall.

Now this is in contrast to what one sees, for example, in the left common coronary
artery of a horse,[12] as illustrated in Figure 7. This is a little more complicated diagram
showing data from a specific animal experiment. The heart has been exposed, and the
velocity profiles have been obtained entering the probe through the outside wall of the
vessel (the wall opposite the myocardium). In puncturing the probe in through the
outside wall, then as one moves across the figure from right to left, one is traversing
from the outside wall to the inside wall. Velocity profiles are presented at various times
through the cardiac cycle, e.g., one fourth of the way, halfway, three fourths of the

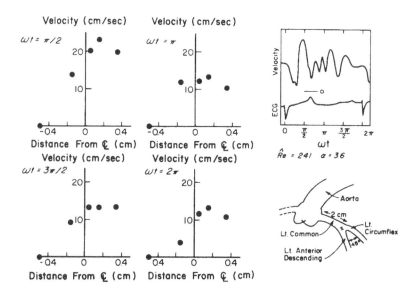

FIGURE 7. Centerline velocity waveform, ECG, velocity profiles for indicated time, and the location of measurement in the left common coronary artery of an anesthetized horse. Measurements were performed with a hot-film anemometer in the plane of curvature of the artery.

way, and full stop as obtained by triggering on the ECG. One can see that, whereas in the dog aorta there was a flat profile skewed due to the curvature, here there is very definitely a fully curved profile with higher velocities toward the outside wall than toward the inside wall. This is the opposite of what was observed in the dog ascending aorta, and there is good reason for this if you look at the flow conditions. For example, if one considers the Reynolds number,* a parameter with which many of you may be familiar, one has a totally different viscous flow condition. The left common coronary artery of a horse has a low Reynolds number, fully viscous flow, whereas the ascending aorta is a high Reynolds number entrance region. In this latter case, there is an inviscid core, and viscous effects are confined to a thin boundary layer (see Figure 4) that doesn't even appear in the profile presented in Figure 6.

Thus, physiologic flows do have asymmetries, and they are not always the same. They are not even always the same in the same vessel. For example, the aortic arch pattern that I discussed was for a dog. However, if one could make detailed measurements in the aortic arch of a rabbit, one would find that the Reynolds number would be sufficiently low so that the flow would be fully viscous. In this case, one might expect the type of skewing shown in Figure 7 as opposed to that in Figure 6.

Another type of complexity of which one should be aware is oscillations with time. Although these may be periodic, relatively pure in frequency, and represent a laminar, unsteady flow, they also may be very chaotic and random, representing turbulence and having the associated broad frequency content. Whether or not turbulence occurs depends on the flow conditions. Do we see turbulence in physiological systems? The answer is yes, at least in certain situations. Figure 8 shows some observations in the ascending aorta of the dog,[13] human[14] (this is a catheter measurement), and horse.[15] These have all been reduced to the same velocity scale, but not the same time scale.

---

* The Reynolds number is defined as velocity multiplied by the characteristic length and divided by the kinematic viscosity. It is a measure of the ratio of the fluid inertia to the viscous force acting on the fluid.

FIGURE 8.   Disturbed-flow velocity waveforms recorded in the dog, human, and horse aorta using a hot-film anemometer.[13-15]

The kind of random, high frequency fluctuations that one thinks of as turbulence are evident. One also observes turbulence in the upper airways, and there are other situations where turbulence might occur which I will not go into here.

I am speaking here about turbulence under normal, physiologic conditions, not pathologic conditions. However, blood flow turbulence also occurs under abnormal, disease conditions such as illustrated in Figure 9. Here a hot-film anemometer measurement of centerline velocity is shown for a position just distal to the point where the descending thoracic aorta of an anesthetized dog has been banded to produce a stenosis.[16] Shown in Figure 9 is the control waveform obtained prior to banding, and the waveforms for the cases of 75 and 95% stenosis. In both these stenosed flow cases, the flow was turbulent as compared to the undisturbed laminar control waveform. Also note that, although the turbulence in the case of 95% stenosis is at least as severe as that for a 75% stenosis, the velocities are much lower.

Figure 10 shows flow visualization in a bifurcation.[17] One can see the dye marking selected flow streamlines. The Reynolds number is 800. In this particular case, the flow is being evenly divided. However, if you look at the dye markers, one of the things to be noted is that they seem to wander around. This is associated with what is called a secondary flow. A flow just does not move from a proximal region to a distal region. It also may have a very definite helical, corkscrew type of motion to it. This type of secondary flow may be very important in trying to understand what is taking place, not only in terms of the physiology, but also in terms of understanding what a given flow instrument output is providing one in terms of information.

Another example of a flow complexity is separation where the flow leaves the wall creating a dead water recirculation region and ultimately reattaches. This is illustrated in Figure 11 for a bifurcation where separation would occur on one side and not the other because of an unequal division of the flow between the two branches. It does not require a torturous geometry to produce flow separation, for the flow may just need to separate in order to accommodate the kind of flow division that is taking place. Obviously, this could influence a flow measurement and one's interpretation of what the flow pattern is. Again, it is the kind of possibility that one should recognize in a physiological flow situation.

Figure 12 illustrates flow separation in a model simulating the in vivo situation where a vein graft has been placed between the iliac arteries of a dog, and the one iliac has been ligated proximal to the graft.[18] This model was constructed to study the use of grafts for bypass surgery. Dye has been used to provide visualization of the streamlines, and as may be seen, there is a large separated region extending far downstream. This separated region is located on the outside wall of what would be the graft and starts at the entrance to the branch or graft just as the flow turns the corner.

There also is the whole question of time variations. Obviously, one can talk about diurnal variations or about those variations associated with the basic heartbeat. However, there may be other types of variations. Figure 13 presents measurements of flow

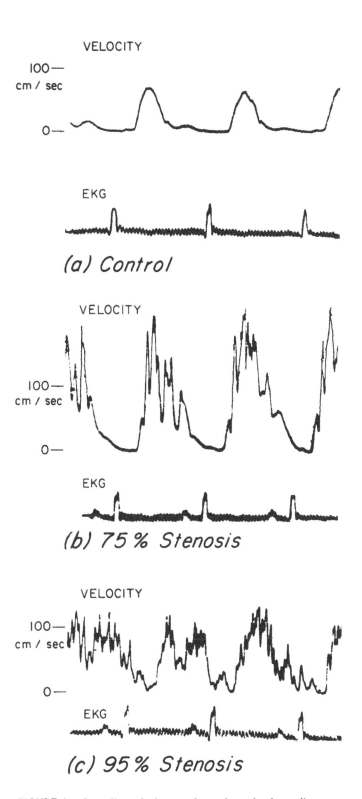

FIGURE 9.  Centerline velocity waveforms from the descending aorta of an anesthetized canine as obtained with a hot-film anemometer for control and stenosed flow conditions where the latter was produced by banding the aorta.

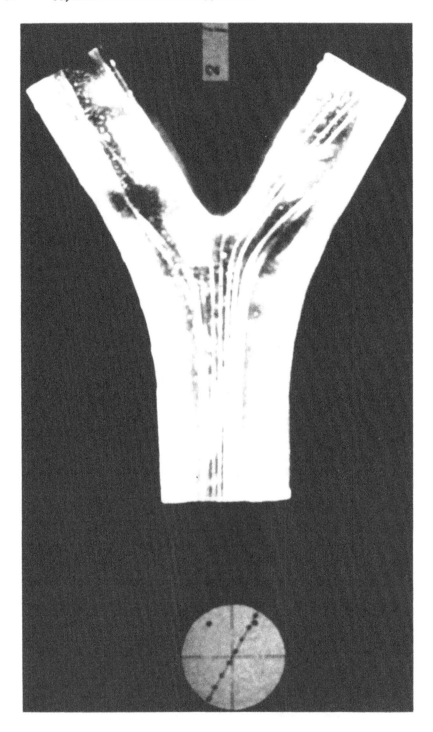

FIGURE 10.   Flow visualization of streamline pattern in a model bifurcation with steady flow at a Reynolds number of 800 and where the flow evenly divides into the two daughter tubes. (From Talukder, N., Untersuchung uber die Stromung in arteriellen Verzweigungen, Dissertation, Technische Hochschule Aaachen, Aachen, West Germany, 1974. With permission).

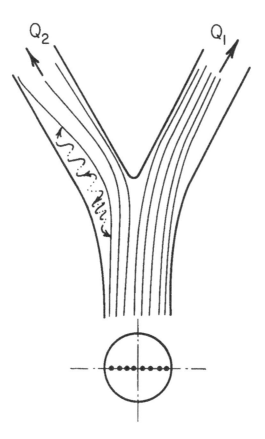

FIGURE 11. Illustration of separated flow occurring in a region of bifurcation where flow into the two daughter tubes has divided unevenly such that $Q_1 > Q_2$.

oscillations as observed in the coronary system of a horse,[12] and it can be seen that, in addition to the basic heart rate as indicated by the ECG, higher frequency oscillations are superimposed. These oscillations have a frequency of 5 to 10 Hz. They appear to be introduced by the fact that in the coronary system during systole the larger coronary vessels lying on the outside of the heart basically represent a closed-off system. This is because the heart muscle is contracted, and there is no place for the blood to go. Thus, a pulse is being injected into a closed-end elastic tube.[19]

These are just some examples of the kind of complexities that can exist physiologically and which, I think, we need to recognize in selecting an instrumentation system and in interpreting what that instrumentation system is providing us as data.

Let us now return to the hot-film anemometer probe of Figure 1 and the high frequency characteristics of this system. Obviously, the electrical circuit of which this film is a part has certain high frequency characteristics. However, the frequency limitation on this system is provided by the fact that the response of this probe is based on a certain relationship of the rate of cooling, the rate of heat loss, and the blood velocity[9,20] (see Figure 1). This relationship is determined by the detailed nature of the boundary layer flow over the tip of the probe. To be able with any ease to interpret the output from this probe, one needs to operate in a quasi-steady range. This places certain limits on the frequency response of the probe, and the high frequency characteristics of the system are a function of a parameter which is frequency, multiplied by

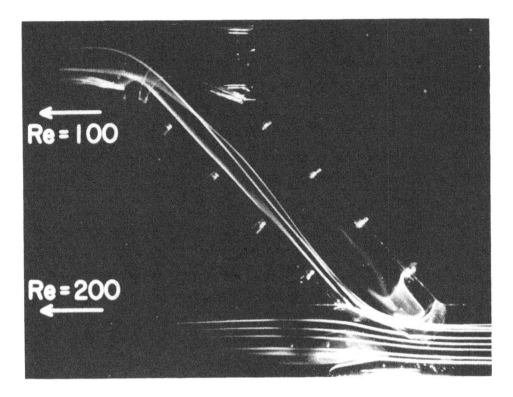

FIGURE 12.    Flow visualization in a steady-flow model of a vein bypass between two iliac arteries, one ligated proximally. Flow Reynolds number is 200 in the parent tube and 100 in each of the daughter tubes.

the distance, x, from the tip of the probe to the film, and divided by the velocity, U.[3,21] If one looks at relative output as a function of this parameter which essentially controls the quasi-steady nature of the boundary layer, one finds that once this parameter becomes too large deviations appear. Such deviations or discrepancies, artifacts if you wish, occur if the frequency is too high, if the distance of the film from the tip of the probe is too large, or if the velocity is too low. Thus, with regard to the latter, this type of device has very poor high frequency characteristics at low velocities and very good high frequency characteristics at high velocities. If one considers the measurement of an aortic velocity waveform where the velocity rises up to its peak systolic value, then falls off, and for a good part of the cycle hovers around zero, one sees that during the diastolic phase one would not have very good frequency response characteristics because the velocity would be low.

As noted earlier, turbulence in a system is one of the complexities in which one might be interested, if not for itself, because of what it might do in terms of the performance of one's instrument. Figure 14 illustrates the use of a hot-film anemometer in a turbulent flow-pipe flow.[12] As the flow was started, the velocity rose and reached a fully developed turbulent flow condition. Then the flow was stopped and slowly decelerated over a 20 sec period. As one can see, the turbulence just doesn't stop or turn off. In fact, it persists for a long time, i.e., it takes a finite amount of time for such disturbances to be damped out.

This is a very artificial situation, but think of inspiratory flow in the lungs. In that situation one has a very highly turbulent flow in the trachea. At some point as one progresses down into successive generations, the Reynolds number becomes sufficiently small so that one would not expect turbulence to be present. However, based

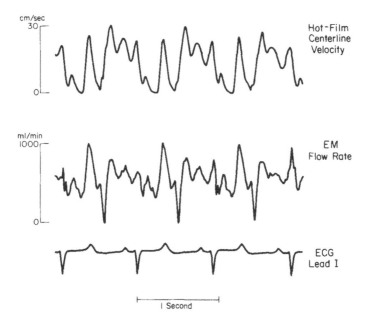

FIGURE 13. Simultaneously measured centerline hot-film velocity waveform and electromagnetic flow meter waveform together with ECG for left anterior descending coronary artery in an anesthetized horse.

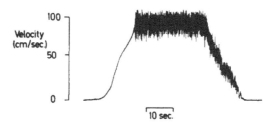

FIGURE 14. Illustration of turbulence in pipe flow where the flow has been suddenly started, maintained constant for a short period of time, and then is suddenly stopped.

on our own measurements, turbulence still is observed for several generations beyond the point where the flow goes subcritical, simply because it takes time for the turbulence to be damped out.

Figure 8 showed hot-film measurements of flow in the ascending aorta, and I next want to discuss the processing of such signals. One of my interests has been turbulence, its spectral content, and how that spectral content changes with time. A method for processing disturbed flow signals is illustrated in Figure 15.[22] In the upper part of the figure, the solid line is the single-beat velocity, i.e., the velocity waveform as measured for a single beat in the ascending aorta. This waveform is part of a series of 270 waveforms. One can do an ensemble average, where one averages on a point by point basis across these 270 cycles and produces an average waveform. This is represented by the dashed line in Figure 15. Turbulence is by definition a chaotic, random type of event and should average out in that type of ensemble averaging process. If one then subtracts the ensemble average from the instantaneous single beat velocity, one obtains

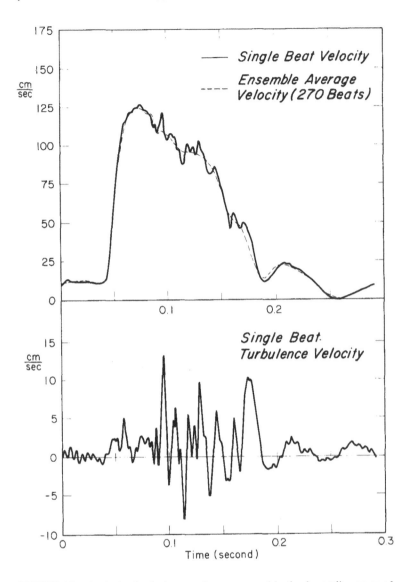

FIGURE 15.   Analysis of velocity waveform measured in the descending aorta of
an anesthetized dog showing single-beat velocity, ensemble average velocity, and
single-beat turbulence or disturbance velocity.[22]

an error signal or what is called here the single-beat turbulence or disturbance velocity.
Observe the turbulent nature of this signal, the fact that there is a variety of frequencies present, and that the amplitude is somewhat random. This disturbance, or single-
beat turbulence velocity, appears to represent the type of chaotic, broad frequency
band signal that one associates with turbulence.

Now, there is one problem here. Although the systolic portion of the single-beat
turbulence velocity is probably turbulence, or something building towards turbulence,
the flow reversal part of the single-beat turbulence velocity really is associated, if anything, with the data processing of the signal. The probe used in this case did not have
a directional capability, and furthermore, with these kinds of low diastolic velocities,
the hot-film has a very difficult time trying to follow exactly the events taking place.
With the kind of differences that appear in terms of where flow reversal actually oc-

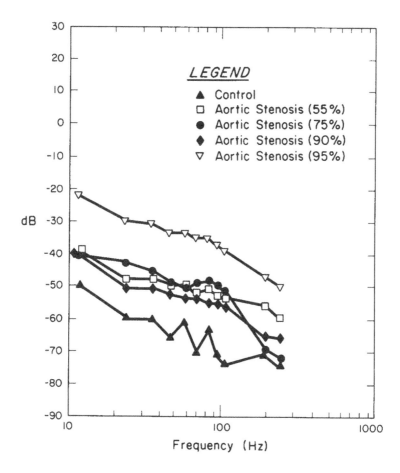

FIGURE 16.   Power spectra of the disturbance signal for centerline velocity waveforms measured in the aortic stenosis experiment of Figure 9. Analysis is for time
segment corresponding to peak systole.

curs, one does obtain a high frequency component that does not average out in the
data processing, but shows up instead in the single-beat turbulence or disturbance velocity as an artifact.

One of my interests is spectral content as a function of time. In Figure 16 are presented power spectra calculated using the method of processing similar to that described above for Figure 15. These power spectra presented in Figure 16 correspond
to the experimental study of stenosed flow[16] from which centerline waveforms were
illustrated in Figure 9. The power spectra are for both the control condition and several
different stenosed flow conditions corresponding to 55, 75, 90, and 95% stenosis. The
method of processing used allows for obtaining time-resolved spectral content, and
the spectra shown are for that time segment corresponding to peak systole. In each
case, the spectrum has been nondimensionalized by the mean velocity for the corresponding time segment. As can be seen for such spectra, even though the peak systolic
velocities are considerably different in each case, there is little difference in the  nondimensionalized distribution of the high frequency waveform content, with two excep-
tions. In the control case, the high frequency content is considerably suppressed, and
in the 95% stenosis case, it is noticeably enhanced.

Let me go on to a slightly different subject related to flow. What I have illustrated
in Figure 17 are the hemodynamic forces that would be exerted on a blood vessel wall.

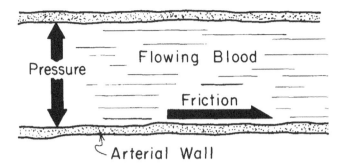

FIGURE 17.   Forces exerted by flowing blood on the arterial wall.

This would be the same for any tube containing a moving fluid in any physiological situation. If you look at the flow over a surface, there is a force imposed by that flow on that surface which has basically two components: the pressure force acting normal to the wall and the shearing stress acting tangential to the wall. Although we may "think" we know how to measure pressure, when it comes to measuring shear stress our ability is extremely limited. For example, although one can measure velocity profiles with the kind of instrumentation I previously discussed, it is virtually impossible to perform such measurements with a detail at the wall sufficient to obtain a good shear-stress estimate from the velocity profile.

There are ways of measuring shear stress directly. One technique involves hot-film anemometry,[1,23] and there also is the electrochemical technique,[24, 25] which is very similar in principle to the hot-film technique. Both of these techniques can be used for shear-stress measurement. Both involve some type of a sensor flush mounted in the surface. In the hot-film case, one is actually measuring heat transfer to the sensor and how flow affects it. The primary determinant of the effect of flow on heat transfer is the shearing rate. For the electrochemical technique, one is looking at the diffusion of a chemical where there is a chemical reaction at the surface and the diffusion is influenced by flow through the shearing rate. So there are these two techniques, very similar, one involving heat transport and one involving mass transport. Both can be used to measure shear stress at a surface, and in both cases it is somewhat difficult to interpret the results if the flow is not steady. Furthermore, it is certainly extremely difficult, if not totally impossible, to think about a flush-mounted probe in a physiological situation.

Thus, although we have some ability to measure flow, one can have no real confidence in our ability to measure wall shear stress. This is unfortunate, for there is considerable interest, at least in certain problem areas, in wall shear stress. For example, in the study of the role of hemodynamic forces in atherogenesis, one of the problems of interest is how wall shear stress might influence the endothelium and transendothelial transport. Unfortunately, one cannot make the kind of in vivo measurement one would wish, and velocity profile information, although suggestive, does not represent the final answer.

In this presentation, it has not been my intent to provide a comprehensive survey of physiological measurement techniques. What I have attempted to do is to provide a look at one aspect of flow measurement — the flow itself. There is an adage which says something about "what you see is what you get". However, what you see depends not only on the instrumentation system, but also on the observer. What we obtain from flow measurements many times is limited not just by the system, but also by the observer. Thus, experimental measurements are not just an instrumentation problem. We also need to understand a little about the quality of the events that we are trying to measure.

Think about some of the complexities that I have here discussed. One of the standard techniques for measuring flow is the electromagnetic flowmeter. If there is an asymmetric flow pattern, which you almost always do have, what does that mean to an electromagnetic flowmeter? If one is trying to look for turbulence, would one even see it with an electromagnetic flowmeter? If there happens to be a flow separation right within the flowmeter itself for some reason, what does that mean in terms of output of the flowmeter?

I don't mean to single out electromagnetic flowmeters. Certainly from an engineering point of view, one tries to select an instrument which will do the job at the level required. Sometimes this means using a very sophisticated approach, sometimes not. However, I do think we need to recognize the limitations of what we do, and one of those limitations involves the complexities of the flow and our perception of those complexities.

## REFERENCES

1. Ling, S. C., Atabek, H. B., Fry, D. L., Patel, D. J., and Janicki, J. S., Application of heated-film velocity and shear probes to hemodynamic studies. *Circ. Res.*, 23, 789, 1968.
2. Schultz, D. L., Tunstall-Pedoe, D. S., Lee, G. de J., Gunning, A. J., and Bellhouse, B. J., Velocity distribution and transition in the arterial system, in *Circulatory and Respiratory Mass Transport,* Wolstenholme, G. E. W. and Knight, J., Eds., J. & A. Churchill, London, 1969, 172.
3. Seed, W. A. and Wood, N. B., Development and evaluation of a hot-film velocity probe for cardiovascular studies, *Cardiovasc. Res.*, 4, 253, 1970a.
4. Baker, D. W., Pulsed ultrasonic doppler blood-flow sensing, *IEEE Trans. Sonics Ultrason.*, SU-17, 170, 1970.
5. Peronneau, P., Hinglais, J., Pellet, M., and Leger, F., Velocimetre sanguin par effet Doppler a emission ultrasonore pursee, *Onde Electr.*, 50, 369, 1970.
6. Histand, M. B., Miller, C. W., and McLeod, F. D., Transcutaneous measurement of blood velocity profiles and flow, *Cardiovasc. Res.*, 7, 703, 1973.
7. Baker D. W. and Daigle, R. E., Noninvasive ultrasonic flowmetry, in *Cardiovascular Flow Dynamics and Measurements*, Hwang, N. H. C. and Normann, N. A., Eds., University Park Press, Baltimore, 1977, 151.
8. King, L. V., On the convection of heat from small cylinders in a stream of fluid: determination of the convection constants of small platinum wires with applications to hot-wire anemometry, *Philos. Trans. R. Soc. London*, A214, 373, 1914.
9. Clark, C., Thin film gauges for fluctuating velocity measurements in blood, *J. Phys. E.*, 7, 548, 1974.
10. Wells, M. K., Bhazat, P. K., and Rumberger, J. A., Comparison of arterial velocity waveforms using pulsed Doppler and hot-film velocity meters, in *Dig. 10th Int. Conf. Med. Biol. Eng.*, Albert, R., Vogt, W., and Helbig, W., Eds., International Federation for Medical and Biological Engineering, Dresden, German Democratic Republic, 1973, 20.
11. Seed, W. A. and Wood, N. B., Velocity patterns in the aorta, *Cardiovasc. Res.*, 5, 319, 1971.
12. Nerem, R. M., Rumberger, J. A., Jr., Gross, D. R., Muir, W. W., and Geiger, G. L., Hot-film coronary artery velocity measurements in horses, *Cardiovasc. Res.*, 10 (3), 301, 1976.
13. Nerem, R. M. and Seed, W. A., In vivo study of the nature of aortic flow disturbances, *Cardiovasc. Res.*, 6, 1, 1972.
14. Seed, W. A. and Thomas, I. R., The application of hot-film anemometry to the measurement of blood flow velocity in man, in *Fluid-Dynamic Measurements in the Industrial and Medical Environments*, Proc. DISA Leicester University Press, 1972, 298.
15. Nerem, R. M., Rumberger, J. A., Jr., Gross, D. A., Hamlin, R. L., and Geiger, G. L., Hot-film anemometer velocity measurements of arterial blood flow in horses, *Circ. Res.*, 34, 193, 1974.
16. Sundararajapuram, S. S., The Analysis of the Spectral Characteristics of Disturbed Blood Flow, Ph.D. dissertation, Ohio State University, Columbus, 1977.

17. Talukder, N., Untersuchung uber die Stromung in arteriellen Verzweigungen, Dissertation, Technische Hochschule Aachen, Aachen, West Germany, 1974.
18. Rittgers, S. E., In Vitro and In Vivo Studies of Fluid Mechanics Within Vein Grafts Used as Arterial Bypasses, Ph.D. dissertation, Ohio State University, Columbus, 1978.
19. Rumberger, J. A. and Nerem, R. M., A method-of-characteristics calculation of coronary blood flow, *J. Fluid Mech.*, 82 (3), 429, 1977.
20. Pedley, T. J., On the forced heat transfer from a hot-film embedded in the wall in two-dimensional unsteady flow, *J. Fluid Mech.*, 55, 329, 1972.
21. Seed, W. A. and Wood, N. B., Use of hot-film probe for cardiovascular studies, *J. Phys. E.*, 3 (Series 2), 377, 1970b.
22. Parker, K. H., Instability in arterial blood flow, in *Cardiovascular Flow Dynamics and Measurements*, Hwang, N. H. C. and Normann, N. A., Eds., University Park Press, Baltimore, 1977, 633.
23. Janicki, J. S., Application of the Heated-Film Shear Probe Technique to the Measurement of Certain Hemodynamic Events, Thesis, Catholic University of America, Washington, D.C., 1970.
24. Lutz, R. J., Cannon, J. S., Bischoff, K. B., Dedrick, R. L., Stiles, R. K., and Fry, D. L., Wall shear stress distribution in a model canine artery during steady flow, *Circ. Res.*, 41 (3), 391, 1977.
25. Smith, D. A., Colton, C. K., and Freedman, R. W., Shear stress measurements at bifurcations, in *Fluid Dynamic Aspects of Arterial Disease*, Nerem, R. M., Ed., Ohio State University, Columbus, 1974, 12.

# DISCUSSION

**Dr. Charles Knapp:** I have a question about the power spectra that you showed and the 270 beats over which you averaged. If those are not exactly the same each time — let's say they are not turbulent necessarily, but just not reproducible — how can you distinguish between what you call turbulence and, possibly, the fact that they are not reproducible?

**Dr. Nerem:** Those waveforms are not totally reproducible. That is something you and I both know, and it is just difficult to compare that spectrum with a more controlled fluid dynamic situation. What we have done is carried out the same type of data processing on signals that do not have the kind of high frequency content that Figure 15 demonstrates. One can calculate the single beat turbulence velocity. One also can calculate the power spectrum as well as an rms turbulence velocity as a function of time (I did not illustrate the latter). The level is down by an order of magnitude in terms of the rms turbulence velocity and in terms of the spectra, but it is a good point.

**Dr. Knapp:** I agree with the concept.

**Dr. Nerem:** There are beat to beat variations, and these will show up in the kind of processing I have discussed.

**Dr. Harold Sandler:** How do you account for disturbances of flow patterns or the effects of the transducer, particularly with the hot-wire anemometer in the flow stream? When you have a transducer in the flow stream the way you do, to make sure that you are not causing significant turbulence or effects just by the transducer, how do you avoid this?

**Dr. Nerem:** There is a disturbance from the transducer which is carried downstream and does not influence the probe. Beyond that, one has a situation where one is using the known boundary layer behavior over that instrument, the known interaction, as our technique for measuring velocity. There is a boundary layer that develops on the probe. That boundary layer has certain characteristics, and those characteristics are important in terms of the actual output of the device. We are using that known "disturbance" (if you wish) as a calibrated way of making a flow velocity measurement. Certainly, when the flow is swept back over the probe the other way, it's a totally different situation. The particular probe I showed is built primarily for systolic velocity measurements.

**Dr. Walter Olson:** In your diagram for the hot-film probe, you showed the sensor right in the middle of the vessel. In vivo, how do you assure yourself that it stays there and, if it doesn't stay there, what effect does it have?

**Dr. Nerem:** I do not have a diagram to show you. However, the way we position these probes is to come in through the vessel wall and essentially fix the distance from the wall the probe enters through to the probe. Thus, as the area changes slightly due to pressure changes, the probe position will move slightly in that plane, but the probe will remain a fixed distance from what I call the near wall.

**Dr. Olson:** It doesn't make a difference where it is then?

**Dr. Nerem:** That, in a sense, is the whole purpose of the measurement, because as you get close to a wall, you are going to have lower velocities than you will have further out. Our interest has been in defining that variation.

**Dr. Donald Baker:** Have you done any work to compare the Doppler spectra to the hot-film spectra, and if so, what is your impression?

**Dr. Nerem:** No, although we are getting close to it because we are building a pulsed Doppler unit in our laboratory so that we can do direct comparisons. I know there has been some limited work done at the University of Washington. Up to a few years ago, I had very little confidence in the ability of the pulsed ultrasonic Doppler to sense turbulence. It seems that the state-of-the-art now has developed beyond what it was at that point. I do see turbulence being sensed by pulsed ultrasonic Doppler units, but I think that as far as any kind of detailed comparison, probably the true spectra will lie somewhere in between the two produced by the two instruments.

**Dr. Olson:** What is the size of the instrument?

**Dr. Nerem:** The diameter of the probe is about 0.8 mm. We build our own probes. A vessel like the left common coronary artery of a horse may be on the order of 8 to 10 mm in diameter. Obviously, blockage effects are important, so you want to minimize probe size as much as possible.

**Alan Furler:** Could you give a little more detail on the fabrication of the thin film probes, materials involved etc.?

**Dr. Nerem:** For the thin films that we make ourselves (we also buy some from DISA and TSI), we deposit a thin platinum film on a quartz or glass substrate and then coat it with some kind of an insulating material. There is the possibility of electrical leakage, although it really doesn't seem to interfere with the output of the probe and, in an animal situation, is not too serious. There have been attempts to go into humans with hot-film catheters. Although my friends in England don't seem to worry about the electrical leakage, in the United States we do. I think that's for obvious reasons.

*State of the Art*

Chapter 4

# TRANSDUCERS FOR IN VIVO MEASUREMENT OF FORCE, STRAIN, AND MOTION

## James B. Agnell

## TABLE OF CONTENTS

# I. INTRODUCTION

In this presentation we will consider two transducers, an accelerometer and a force transducer, which serve as examples of a new field which has been labeled "micromachining". Micromachining is the combination of conventional silicon integrated-circuit technology, the process by which regular integrated circuits are made, together with advanced chemical etching techniques which enable us to remove silicon in a very precise, controllable, reproducible manner. With this combination of technologies, it is possible to accurately and economically make carefully designed precision structures on a batch fabrication basis. Thus, we hope to extend to transducer fabrication the economies which IC technology has given us in circuit fabrication.

# II. ACCELEROMETER

The first structure we will consider is an accelerometer.[1] It is the newer to the two structures to be described and represents a very elegant application of the micromachining technology. The original goal for the accelerometer development was a biomedical application. Therefore, it was important that it be as small and lightweight as possible in order to avoid measurement error resulting from loading by the transducer. For an accelerometer, which measures acceleration in one particular direction (three transducers could be used if three-axis measurements are needed), it is necessary to have the package give a good reference with respect to the sensitive axis of the accelerometer.

Figure 1 is a sketch of the accelerometer. It is, in principle, a very simple structure. There is a cantilevered beam of silicon with a silicon mass on its free end. As the structure is accelerated, the mass will cause the beam to bend. The mass itself is a nearly square paddle of silicon. In the top surface of the beam is a U-shaped resistor of p-type silicon. This p-type resistor is piezoresistive, so that its resistance changes (from a nominal starting value of 5000 $\Omega$) in direct proportion to the stress in the upper surface of the beam. The stress itself is directly proportional to acceleration. The resistance change is measured through leads attached to the bonding pads. Because silicon resistors also change with temperature, we have included a second resistor, identical in size and also made on thin silicon with the same thermal inertia as the beam, as a thermal reference. Therefore, variations with temperature can be externally compensated.

The starting silicon wafer is 200 $\mu$m thick and is etched from the bottom side, through a pattern which defines the beam, mass, and surrounding rim, down to a thickness of 10 to 20 $\mu$m. This etch also defines the outer dimension of the transducer and forms the lanes along which the wafer will later be sawn into individual transducers. A shallow etch from the top of the wafer forms the airgap around the mass and completes the formation of the beam. All the etching is done with a solution of potassium hydroxide, which is an anisotropic etch for silicon. This anisotropic etch will, at proper concentration and temperature, etch the (100) planes of silicon 100 times as fast as it etches the (111) planes. Because the starting silicon is a (100) wafer, the etch tends to form V-grooves. The exposed (111) planes are clearly visible in the scanning electron micrograph shown in Figure 2.

One important reason for using an anisotropic etch is the fact that such etches will not undercut the photolithographically defined etch-mask pattern in the SiO$_2$ on each surface of the silicon wafer. Thus, very precise geometrical patterns can be formed.

The top and bottom glass covers begin as glass wafers at least as large as the starting silicon wafer. Cavities are etched in both glass wafers so that the mass of each accelerometer will have room in which to move.

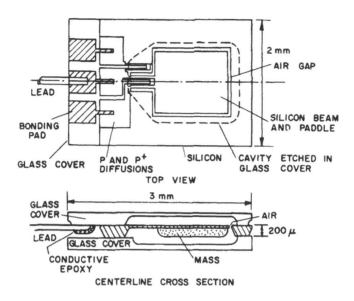

FIGURE 1. Top and cross-section views of accelerometer.

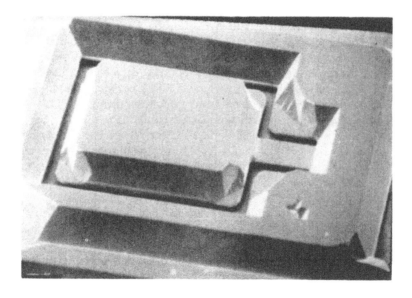

FIGURE 2. Scanning electron micrograph of the accelerometer bottom.

These cavities in the glass caps will provide a mechanical stop so that the mass will hit a limit before the beam itself will fracture. Silicon which is 10 $\mu$m thick is quite flexible and can withstand considerable bending before it breaks. A metal pattern is also defined on the upper glass cap in order to provide bonding pads for lead attachment.

The glass caps (in wafer form) are attached, one at a time, to the silicon wafer by an extremely attractive process called anodic bonding.[2] Anodic bonding forms an atomic bond (of which the detailed nature is not yet understood) between the silicon wafer and either the silicon or silicon dioxide in the glass. It is accomplished by sandwiching the glass between the silicon (after they are carefully aligned) and a metal

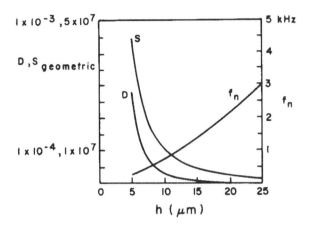

FIGURE 3.   Accelerometer sensitivity (S), resonant frequency
($f_n$), and droop (D) vs. beam thickness.

plate. The sandwich is heated to 300 to 400°C, and the silicon is made 600 V positive
with respect to the metal plate. With the right kind of glass (for example, Pyrex®),
the sodium ions within the glass become mobile at elevated temperature and tend to
migrate away from the silicon. A thin layer which is depleted of mobile ions forms
right next to the silicon, and after a few minutes, all of the applied 600 V appears
across the layer. This high electric field gives a strong force of attraction between the
silicon and glass (around 50 psi) and pulls them into very intimate contact. At the same
time, the high field helps form a good electronic bond between the atoms in the silicon
and glass, giving a strong, irreversible hermetic bond in just a few minutes.

After both glass covers are bonded, we have a 2-in. sandwich containing 160 accelerometers. Now the wafer is sawn apart using a diamond dicing saw (as is often used
for saw-dicing silicon integrated circuits or discrete components).

One of the most important quantities affecting the sensitivity of the accelerometer
is the thickness of the beam, h. In Figure 3 we show how three quantities vary with h.
For usable beam thickness (greater than 8 μm), the resonant frequency varies from 0.7
to 3 kHz. The sensitivity varies almost inversely with the resonant frequency, so that
their product is a constant.

For the most sensitive device (thinnest beam feasible) a sensitivity of 0.4% change
in resistance per g of acceleration is obtained. With such a device, accelerations from
0.001 to 10 g can be measured. One quantity which we hadn't originally envisioned as
important is the "droop", or the motion of the tip of the mass for 1 g (its own weight).
As Figure 3 shows, the droop increases much faster than sensitivity and says that the
beam should not be thinner than 8 μm.

Transient response tests show that the typical accelerometer has a high-Q resonance
with a Q of 80, which would tend to limit the frequency response to perhaps 20% of
the resonant frequency. Another disadvantage of the resonance is the possibility of
breaking the beam if the accelerometer experiences a mechanical shock impulse. It was
found that by filling the cavity with isopropyl alcohol (through an etch hole which is
later sealed with epoxy) close to critical damping can be achieved, thus improving the
measurement bandwidth and resistance to shock.

Leads are attached to the bonding pads on the glass with a conductive silver epoxy.
The space between the two glass caps is then filled with a nonconductive epoxy for
added strength at the leads. A heavy coat of Parylene®* can be used in order to
provide a moisture barrier which will survive many months in vivo.

---

*   Parylene® is a proprietary polymer coating developed by Union Carbide Company.

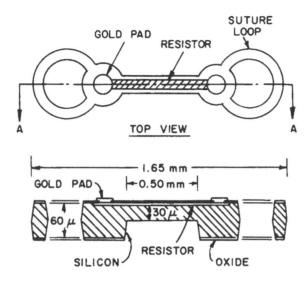

FIGURE 4.    Force transducer.

## III. FORCE TRANSDUCER

A micromachined silicon force transducer, originally developed for in vivo biomedical instrumentation in 1972, is shown in Figure 4. It is a 1.7-mm long piece of silicon with a stress-sensitive resistor diffused into the narrow middle section. At each end of the resistor are gold bonding pads which make ohmic contacts to the resistor. The silicon ring at each end of the transducer serves as a suture loop for surgical implantation. In the cross-section view of Figure 4, the vertical scale has been expanded 4:1 for clarity.

The fabrication of these devices begins with a silicon wafer that has been chemically thinned to 60 $\mu$m. The processing steps are similar to those for the accelerometer and use KOH for the anisotropic etching. The silicon under the piezoresistor can be thinned to 30 $\mu$m (for greater sensitivity) by permitting the KOH to attack only the bottom surface of the silicon at that point, whereas the etchant attacks both top and bottom surfaces to form the suture loops and the perimeter of the device. When the etching of a given device is completed, the device separates from the wafer, falls to the bottom of the etching bath, and is removed. Over 1000 working transducers can normally be obtained from a 2-in. wafer.

These devices are surprisingly strong. They can withstand 90 g of tension or 0.5 g·cm of bending moment before fracturing. At 10% of the breaking stress, the resistor (nominally 1100 $\Omega$ has changed by 3%. Up to this point, the resistance change is proportional to stress with no sign of hysteresis.

Figure 5 shows a photograph of a completed force transducer. Note that the suture loops, which began as circular sectors on the masks, have been distorted by the directional etching characteristics of the anisotropic etchant.

This structure is certainly temperature dependent, and we do not have a temperature compensating resistor on it. Thus, it is necessary to make the device sensitive enough that the small temperature changes within tissue are not significant in terms of the resistance change due to the forces that are being measured.

A sensor with two platinum-iridium leads silver epoxied to the bonding pads is shown in Figure 6. At a distance of 2 cm from the transducer, the leads, which are both flexible and very strong, are soldered to multistranded copper wires which are

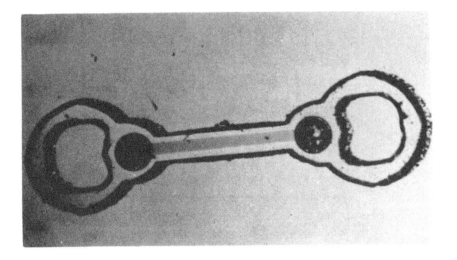

FIGURE 5.   Photomicrograph of a force transducer.

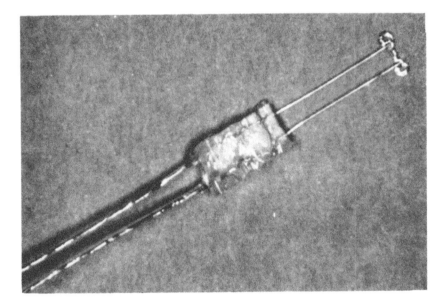

FIGURE 6.   Force transducer assembly.

used, in part, to keep the series resistance low. (If we were using implanted telemetry instead of hard wiring, the copper wires would not likely be needed.) The soldered joint is embedded inside a silastic block by making two holes in the block with needles and then inserting the platinum-iridium leads behind the needles as they are withdrawn. A suture hole is made for stitching the block down to some adjacent tissue in order to provide strain relief and keep forces that might be transmitted through the lead wires due to tissue motion from reaching the transducer. The multistranded copper leads, which are insulated with Teflon®, are enclosed in a plastic sleeve for a distance of 10 cm, and the entire structure is coated with Parylene® of the order of 10 $\mu$m thickness. This Parylene® coating makes the lead structure much stronger, as well as providing moisture and corrosion protection for at least 70 days.

## IV. SUMMARY

These, then, are two examples of using silicon technology for micromachining to shape silicon into desired forms. We are obtaining experience with other applications of the same set of "micromachining" tools for such structures as pressure transducers, arrays of pressure transducers, temperature-sensor arrays, and miniature gas chromatographs. We are thus attempting to study and develop technology which will extend the advantages of batch integrated-circuit fabrication into the world of sensors.

## ACKNOWLEDGMENT

This silicon transducer research has been supported by the United States Government through Grant No. NGR 05-020-690 from the National Aeronautics and Space Administration and through Contract No. NOI-HD-3-2774 and Grant No. 1-PR1-RR01086 with the National Institutes of Health.

## REFERENCES

1. **Roylance, Lynn M.**, A miniature IC accelerometer, *IEEE Int. Solid State Circuits Conf. Dig.*, Lewis Winner, Coral Gables, Fla., 220, 1978.
2. **Wallis, G. and Pomerantz, D. I.**, Field assisted glass-metal sealing, *J. Appl. Phys.*, 40, 3946, 1969.

## DISCUSSION

**Anon.:** What are some of the uses of these devices?

**Dr. Angell:** Dr. Nelsen will be describing applications of the force sensor. I can't tell you very much about the use of the accelerometer because we haven't actually used it in any in vivo experiments, but I claim that it is suitable for in vivo use because it is small and it is Parylene® coated. We don't have to worry about any measurement artifice introduced by the Parylene® because it doesn't reach the beam.

**Dr. Hunter Peckham:** I'm interested in the packaging of the accelerometer and what limitations there might be with the temperature, voltage, pressure bonding technique in terms of the devices. What has your experience been?

**Dr. Angell:** Yes, we have had a lot of experience with anodic bonding work. We've been at it for about 6 years. There are a couple of characteristics of the glass that are necessary. One obvious one is that it have a temperature coefficient of expansion fairly similar to that of silicon, for obvious reasons. The second is that it has to be a glass that has enough sodium in it to provide a slight conductivity at the temperatures at which we do the bonding, which are typically 300 to 400°C. The glass that we have found most successful for this is a Corning glass #7740. It is Pyrex® glass whose temperature coefficient of expansion is within 10% of that of silicon, which seems to be good enough. We first used it in making column gas chromatography system on a wafer, not on a chip, but a wafer 1½ to 2 in. in diameter. We have an 8-m-long column which has been etched as a phonograph groove. The glass is then bonded onto the top of that phonograph groove, where we have left enough unetched material between grooves so the glass can bond to it. It has become a routine procedure with less than one miss in twenty, and they never break. Apparently, that coefficient of expansion match is close enough so that it doesn't give any significant strain problems into the silicon which would cause it to fracture.

**Dr. Peckham:** What sort of time is involved?

**Dr. Angell:** It's a marvelous thing to watch. The bonding in any one area takes place, probably, in 10 to 15 sec. As one region has made contact, it converts into a sort of greyish color. Then you can watch, by the color fringes, as the bonded area spreads out from that point; it normally spreads over the entire wafer within 2 min. If there is a little dust particle, you can see the little circular region surrounding it that is not bonded. Nevertheless, at 350°C we normally maintain the bonding process for 10 to 15 min with the voltage on, and then the bonding is complete. It's a patented process conceived by chaps named Wallis and Pomerantz somewhere in the late 60s, (at least that is when they published), and apparently that same process is in commercial use for a variety of techniques. It serves also for bonding glass to metal as well, not just silicon.

**Dr. Richard Cobbold:** I was fascinated to listen to your presentation, but one question has occurred to me concerning the force transducer in regard to the temperature sensitivity. You have this single resistor on each element and, therefore, do you get a photosensitivity change?

**Dr. Angell:** The sensitivity will change with temperature, and of course, the resistance changes with temperature. There is one doping level at which the resistance doesn't vary with temperature. At very low doping, the temperature coefficient of resistivity goes in one direction, and at a high doping level, it goes in the opposite direction because of the different scattering mechanisms that are involved in the resistivity. At a surface resistivity of somewhere around $10^{17}$, they almost, or do, cancel. Because of the application that Dr. Nelsen has been involved in, temperature was not one of the major problems that we had. He didn't insist that we shoot for exactly the zero temperature coefficient, although the diffusion which we put in the p-resistor approached it. We haven't studied the effect of illumination on the force transducer.

**Dr. Cobbold:** At a resistivity of around $10^{17}$, do I understand that the sensitivity remains constant? I realize that the temperature coefficient of resistance is essentially zero.

**Dr. Angell:** The piezoresistive coefficient itself is not as high as it should be at some other doping level, but it is still good enough to read something with it. It's down by a factor of maybe two or three from what the optimum value would be. The piezoresistive coefficient is somewhat temperature dependent at this doping level, so the sensitivity would change slightly.

**Dr. James Fordemwalt:** Are you using diffused or ion-implanted resistors to get the DC down?

**Dr. Angell:** At this point, both the force transducer and the accelerometer use only diffused resistors simply because we found that they were adequate for these jobs. In the case of pressure transducer work, we had looked very seriously and in fact made some pressure transducers using ion-implanted resistors. The principal advantage is that you can get about 100 times the ohms per square, hence 100 times the resistance in a given geometry. There, the chief virtue is that for a given voltage, say a 1 V power supply across the bridge, you only have a short circuit current of about 10 $\mu$A, which is very biocompatible. In the long run, we would like to think that these things could conceivably be suitable for human use, and in that case of course, you want to make sure that the worst thing that could go wrong would not give rise to any lethal currents. From numbers which I have seen, 10 to 30 $\mu$A is never dangerous. That would be the chief advantage of going to higher resistances. Of course, if you are driving it from an internal power supply, you will save a lot of power by using 100 k$\Omega$ resistors rather than the 5 k$\Omega$ resistors we standardly use in our pressure transducers or in the accelerometer.

**Dr. James Topich:** Would you consider using capacitance changes in the accelerometer to increase its sensitivity?

**Dr. Angell:** In our original work along this line on the pressure transducer that was supported by NASA-Ames, we were deliberately looking for the piezoresistance structure because some of the NASA team with whom we were working did have a capacitive pressure transducer under development. In the case of the accelerometer, the first one was actually one of the force transducers with a big gold ball glued on the end of it, and the whole thing was stuck inside a TO-5 package. It seemed to give sufficiently adequate sensitivity. We felt that the problems of trying to make a capacitive accelerometer, and that you'd have to use a 1 MHz rather than a DC type of sensing system, didn't seem to offer enough advantage. Hence, we never considered it any further than being aware of the fact that it offered an alternative method. We are quite some believers in the piezoresistive coefficient. There's something nice about it. Certainly, you have to understand it. For example, the orientation is best in a (110) direction where it is most sensitive. We have a (100) wafer, but we orient the resistors in the (110) direction within the (100) plane.

Chapter 5

RECENT DEVELOPMENTS IN PRESSURE MEASUREMENTS

Wen H. Ko

TABLE OF CONTENTS

## I. INTRODUCTION

In December 1975, the Biomedical Electronics Resource at Case Western Reserve organized its first workshop on indwelling and implantable pressure transducers. Since then, many new approaches have been used for pressure transducer design. It is thought that it would be desirable to review the principles and performances of those pressure transducers that were not covered in the previous workshop and to summarize the further development of the research reported in the proceedings of the last workshop.

## II. PRESENTATION

In the last few years, a new principle for pressure sensing has emerged which is the time delay of surface acoustic waves on crystalline material.[1-4] If one sets up an acoustic wave propagating across a surface or a shallow channel on a material that can change its elastic property upon stress, then the delay time for propagating a surface acoustic wave between two given points will vary with the stress. This occurs in two ways (1) when the diaphragm is bent, the length between the points is changed, and (2) the propagating velocity of the acoustic wave changes, thereby changing the total delay time. If one incorporates this delay time into the feedback circuit of an amplifier with proper phase relationship, then an oscillator is made. The oscillating frequency will vary if the material is bent under pressure or stress. Figure 1 shows a silicon or quartz material containing an interdigitated transducer at point A, that translates electrical signals to acoustic waves. These propagate through this diaphragm and are detected at point B by another interdigitated transducer to convert the acoustic waves back to electrical signals. This signal is, then, fed back to the amplifier to form an oscillator. Figure 1b shows that the etched diaphragm area will bend when pressure is applied on the diaphragm. Therefore, the oscillating frequency will vary with the pressure. The basic equations are given as follows:[1]

$$f(P,T) \simeq \frac{n}{\tau(P,T)}$$

$$\tau = \frac{\ell}{v}$$

$$\frac{d\tau}{\tau} = d_P \, dP + d_T \, dT$$

$$d_P = \frac{1}{\ell} \frac{\partial \ell}{\partial P} - \frac{1}{v} \frac{\partial v}{\partial P}$$

$$f(P,T) \simeq f_o \left[ 1 - d_{Po} P - d_{To} (T - T_o) \right]$$

where f is the frequency; P, pressure; T, temperature; $\tau$, delay time.

The pressure will affect either the length, $\ell$, or the velocity, V. Thus, the delay time, $\tau$, will change. Unfortunately, when temperature varies, it also affects $\tau$.

a) TOP

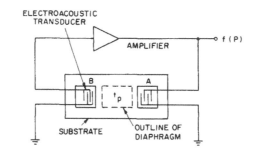

b) SIDE

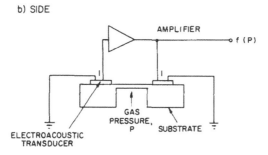

FIGURE 1. Schematic view of the Saw diaphragm pressure sensor.

Several temperature compensation schemes have been developed. One of these is illustrated in Figure 2 where identical surface acoustic delay lines are built on a wafer, and one line is subject to external pressure while the other is not. Since they are made adjacent to each other on the same crystalline wafer, their temperatures track. Let one delay line oscillate at one frequency and the other at another frequency, and the frequency difference of these two oscillators is used as the output. The temperature variation will then be canceled at the output. The signal will then be proportional to the pressure change alone. Such a system has the advantages of (1) the output is in digital form, and (2) the crystal properties and the surface acoustic wave are very stable in time and can be used as a frequency standard, provided temperature is constant.

It is reported that by using silicon (111) wafers bonded on a predrilled substrate to provide a thick base as diaphragms, a diaphragm with a 2 mil thickness, and a 250 mil diameter, will yield a $\tau$ value of 2 $\mu$s. An oscillator oscillating at 167 MHz can have a pressure sensitivity of 0.5 kHz/psi. With the compensation scheme discussed before, a temperature coefficient of 0.05%/°C can be obtained. This is very good as compared to the current piezoresistive device in uncompensated form, or even when compensated with a resistor.

There are several other methods for designing surface acoustic wave pressure transducers, including phase measurement of delayed waves and the use of an acoustic cavity. The latter is illustrated in Figure 3.[1] In an acoustic cavity, a resonator is defined by two parallel reflectors that bounce the waves back and forth. The output is coupled to the electrical circuitry by the middle interdigital transducers. Exploratory results show that for 300 mmHg full-scale pressure transducers, 0.02 mmHg resolution and 0.07% drift is obtained.

Another interesting paper used the GaSb tunnel diode (first reported in 1975 as done at Case Western Reserve University and NASA) incorporated into an oscillator circuit

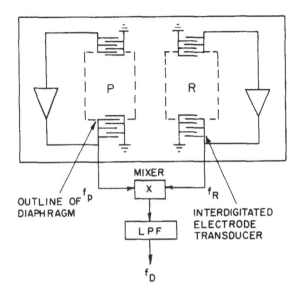

FIGURE 2.  Temperature compensated Saw pressure transducer.

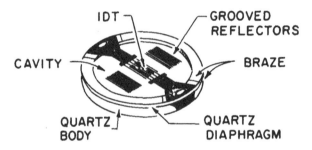

FIGURE 3.    Pressure transducer using Saw resonator.

to make the frequency pressure sensitive.[5] Several other phenomena have been reported, including the deflection of acoustic or optical waves through a diaphragm or air bubble. They will be reported in other articles of this workshop.[6-8]

Next, I would like to summarize the recent developments at Case Western Reserve University.

## A. High Sensitivity Pressure Transducers

These are used for the measurement of venous and other body fluid pressures. The typical range is 10 to 50 mmHg, full scale. At this level, the piezoresistive device will be difficult to maintain the required long-term stable baseline. We are designing a capacitive device to meet this need. Figure 4 illustrates the structure of the device. There is a base plate of the capacitor on one side, and a diaphragm made of silicon or other crystalline material forms the moving plate of the capacitor. When pressure varies, the capacitance changes because of the deflection of the diaphragm. Such a device will be inherently more stable than a piezoresistive device. A comparison of the two types of pressure transducers are summarized in Table 1. In the capacitive device, it is the integrated total capacitance giving a measurement of pressure; therefore, any differential stress set up by temperature or sealing will have a minimal effect, thereby

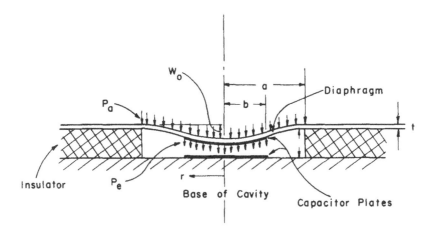

FIGURE 4.  The structure of a capacitive pressure transducer.

TABLE 1

Comparison of Capacitive and Piezoresistive Transducers

| Characteristics | Capacitive device | Piezoresistive device |
| --- | --- | --- |
| Measured parameter | Integrated deflection | Stress on bridge element |
| Usable range | $\Delta C = 1 \sim 10\% \ C$ | $\Delta R = 0.1 \sim 0.5\% \ R$ |
| Full scale output | 10 to 100 mV | $1 \sim 10 \ \text{mV}$ |
| Lateral stress effects | Negligible | Sensitive to stress set up by any causes |
| Effect of non-uniform thermal stress | Negligible | Sensitive to thermally induced stress |

greatly reducing the problems of packaging. A computer analysis of the capacitive transducer was made that provided the expected performance as well as design guides.[9]

Several large bench models were built with two pieces of conductive plate, one silicon wafer with two spacers and a clamping housing. The test results verified the expectation. It was found that sensitivity is good, and over a period of 2 weeks in a laboratory temperature environment, the stability was better than ±1% full-scale. The full-scale pressure was ±20 mmHg. Models 1/2 in.² made of silicon diaphragms electrostatically bonded to a Pyrex base have also been made, and the pressure sensitivity agreed well with the calculated values. Figure 5 shows the stability and sensitivity data of the large models. The design process has begun on an integrated circuit capacitive pressure transducer where one of the on-chip capacitances will be changing with external pressure, another will be for reference, and MOS circuits will be used to excite the impedance bridge circuit and to process the output signal. Our goal is 10 to 50 mmHg, full scale, with a baseline stability of 0.5 mmHg/month in a controlled environment.

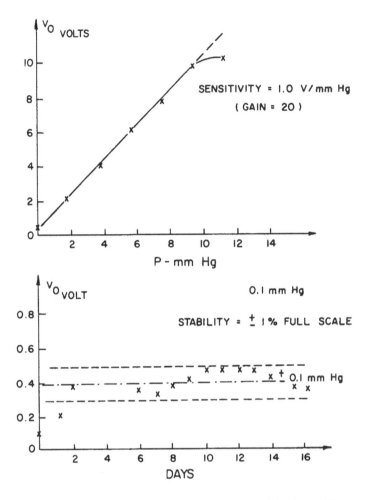

FIGURE 5.   Stability and sensitivity data of large models of capacitive pressure transducers.

## B. Stability of Piezoresistive Transducers[10]

In order to use these devices for implantation, time stability is a very important characteristic. The stability of commercial devices as expressed in specification sheets is generally unacceptable. We have tested several groups of our devices over a period from 6 months to a year.[11] These devices were completed transducers, but were not packaged for biological implantation. Figure 6 shows the structure of the pressure transducer. There are two identical silicon wafers bonded together with a metal alloy. The sensor is a resistor bridge located at the center of the diaphragm with four external leads. There is a cavity which is evacuated to low vacuum to reduce the temperature coefficient. The performance data from over 200 devices tested are summarized in Table 2.

The long-term stability tests indicate that the silicon diaphragm is stable to better than 0.1%/month in baseline. It is the fabrication process and packaging materials that introduced the baseline drift observed.

Devices sealed together with gold-tin alloy consistently showed an aging property when power was applied, as illustrated in Figure 7. As shown here, (1) the diaphragm before assembly is very stable, (2) the diaphragm with sealing compound on it showed a time drift in the negative direction, and (3) when two pieces of diaphragm are put

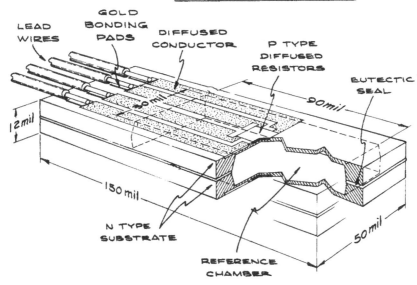

FIGURE 6. Structure of the Piezoresistive pressure transducer.

TABLE 2

**Typical Performance of Resistance Bridge Transducer**

| Characteristics | Sensor | Yield (%) | Diaphragm |
|---|---|---|---|
| Sensitivity ($\mu V$/mmHg-$v$) | $15 \pm 5$ | 70 | — |
| Temperature coefficient (mmHg/°C) | $\pm 1$ | 70 | 0.3 |
| Temperature hysteresis (mmHg/20°C)[a] | 2.0 | 65 | 0.1 |
| Pressure hysteresis (mmHg/100 mmHg) | 0.2 | 70 | 0 |
| Offset voltage (m$V$)[b] | $\pm 20$ mV | 70 | $\pm 20$ mV |

*Note:* Percentage yield of devices satisfying all criteria — 40 to 60%.

[a]  Uncompensated.
[b]  With vacuum chamber reference.

together with sealing alloy, the drift is in the positive direction. The aging-type drift is characterized by a large initial drift followed by a near exponential path to a stable state. Of the devices tested, 80 to 90% showed the aging characteristic; the remaining devices did not stabilize with time. More than 10 devices reached a 1 mm/month stability after aging over the 60 days. The stabilizing or aging effect occurs only when the device is powered. Annealing at 200 to 300°C and cold storage does not improve stability. Some devices showed sensitivity to humidity changes in a complex way. Electrostatic bonding was tried to seal a silicon diaphragm to a #7740 Pyrex® base plate at 400 to 500°C. Early indications are, that for electrostatically sealed devices, there is no aging property, but rather, a nearly linear drift with time. From these data, we concluded that: (1) additional surface passivation methods should be incorporated to eliminate the humidity effect, and (2) a study needs to be conducted to examine why only the devices that are powered demonstrate the gradually improved stability. This

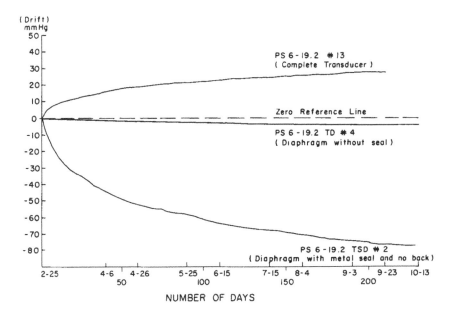

FIGURE 7.    Aging effect of Piezoresistive pressure transducer.

may be related to the properties of the oxide or the surface of the silicon. The monitoring of leakage current and the interface charge may be important clues. The sealing alloy structure must also be explored. For practical biomedical applications, packaging is a major unsolved problem. The present device has reached a good stability (2 mm/ month) with aging under power, but still needs packaging before application. For short-term indwelling applications, the device has been packaged with a polymer coating that can be used over a period of days. For a chronic implant, no practical solution has been found except an additional outer shell that interfaces the device with the body.

For the intracranial pressure monitoring of head trauma or hydrocephalus, intracranial fluid pressure readings may be desired over a period greater than a few days. In such cases, the catheter type of external monitor becomes unacceptable. An intracranial pressure and temperature telemetry system was designed and reported at the last workshop.[12] The piezoresistive diaphragm is used as the transducer. A 3.5 MHz RF signal is used to supply power to the electronic circuits. The unit transmits two channels of signal, pressure and temperature, with pulse position modulation at a carrier frequency of about 120 to 150 MHz. All the hybrid circuits are housed in a 1/2 × 1/2 in. flatpack, and the pressure is coupled in through tubing connected to the shunt bypass connected to a Rickham Reservoir. The implant unit has dimensions of about 3.0 × 1.8 × 0.6 cm and weighs 9 gm. Figure 8 shows a demodulation unit providing a cable to an external coil to supply power to the implanted unit and, at the same time, pick up the ICP and temperature information which is transmitted and processed as indicated on the meter.

Two of the units have been implanted in a dog for 6 and 9 weeks. Experiments done by administering diamol show that there is a drop in the ICP while diamol is injected, with recovery to normal at the termination of the drug. There is a little delay in response and a little delay in recovery time, as shown in Figure 9. The stability over a period of 100 days is about 3 mmHg. One of the systems has been used in a patient as shown in Figure 10.* The boy, who contracted encephalitis, needed the device to de-

* Figure 10 appears on the color insert that follows page 88.

FIGURE 8.    Dimodulator unit of ICP telemetry system.

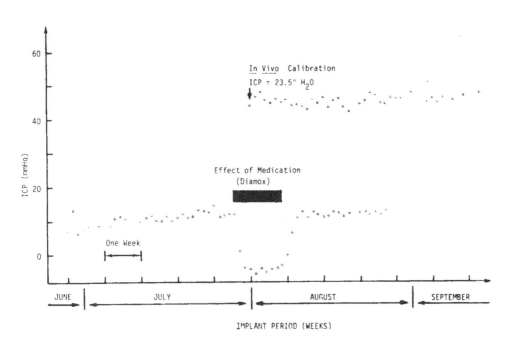

FIGURE 9.    Diamox study and ICP system implanted in two dogs.

termine if he was shunt-dependent so that a clinical decision could be made. The transmitter was taped on the shunt. Continuous recordings were made for several days while he was in the hospital. Once or twice a week recordings were taken. Figure 11 shows the results obtained with the valve on the shunt closed and opened. From this data, the decision was made that he was shunt-dependent. Some other very interesting recordings were obtained, including the patterns of ICP during sleep, the development

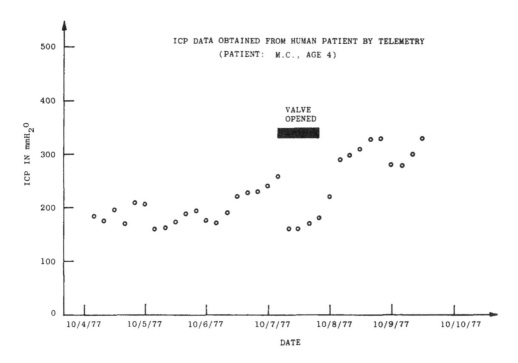

FIGURE 11.    Results of ICP vs. valve position.

leading to the formation of large plateau waves, and the various ICP waveforms while the person resumes normal activity. For this patient, there was a clear pattern of building up plateau waves during sleep which reached a pressure over 60 mmHg. The ICP waveform while the patient was laughing, taking deep breaths, rapid breaths, etc., and at various positions of the head in relation to the body, were recorded. These will be analyzed.

Even though there are many problems still to be ironed out, it is apparent that there are some applications of the present pressure transducers at this time, and we are making serious efforts to look at the time instability to discern the causes and the possible solutions.

# REFERENCES

1. **Reeder, T. M., Cullen, D. E., and Gilden, M.**, SAW oscillator pressure sensors, 1975 Ultrasonics Symposium Proceedings, IEEE Cat. #75 CHO 994-4SU, IEEE, New York, 1976.
2. **Kanda, Yozo and Gross, Chris,** A new type of pressure sensor utilizing surface acoustic waves, *Ferroelectrics*, 10, 71, 1976.
3. **Gilden, M., Reeder, T. M., and de Maria, A. J.,** The mode-locked saw oscillator, 1975 Ultrasonics Symposium Proceedings, IEEE Cat. #75 CHO 994-4SU, IEEE, New York, 1976.
4. **Dias, J. Fleming, Karrer, H. Edward, Kusters, John A., and Adams, Charles A.,** Frequency/stress sensitivity of S.A.W. resonators, *Electron. Lett.* 12(22), 581, 1976.
5. **Ali-Zade, D. G., Leshchinskii, Yu. B., and Ali-Zade, Ya. G.,** Pressure-frequency transducer based on a relaxation oscillator with a tunnel diode as the sensing element (trans.), *Prib. Tekh. Eksp.,* 3, 201, 1976.
6. **Fairbank, William M., Jr. and Scully, Marlan O.,** A new noninvasive technique for cardiac pressure measurement: resonant scattering of ultrasound from bubbles, *IEEE Trans. Biomed. Eng.,* BME-24 (2), 107, 1977.
7. **Cassanta, J. M. and Droms, C. R.,** Miniature low range differential pressure transducer with time constant for R/Vs, presented at Instrument Soc. America 22nd Int. Instrumentation Symp., San Diego, 1976.
8. **Ara, K. and Brakas, M. J.,** Inverse magnetostrictive sensitivity of martensitic stainless steel AISI-410 and its application to pressure measurements, *IEEE Trans. Magn.,* Mag-11(5), 1352, 1975.
9. **Grill, T.,** The Design of an Integrated Circuit Capacitive Pressure Transducer, M. S. thesis, Case Western Reserve University, Cleveland, 1978.
10. **Ko, W. H., Hynecek, J., and Boettcher, S.,** Implantable pressure transducer for biomedical applications, presented at 27th Electronic Component Conf., Arlington, Virginia, May 16 to 18, 1977.
11. **Boettcher, S.,** Stability Test, EDC Memo #226, Engineering Design Center, Case Western Reserve University, Cleveland.
12. **Lorig, R. J., Cheng, E. M., and Ko, W. H.,** Systems for the long-term monitoring of intraventricular pressure in neurosurgery, in *Indwelling and Implantable Pressure Transducers,* Fleming, D. G., Ko, W. H., and Neuman, M. R., Eds., CRC Press, Cleveland, 1977.

Chapter 6

# A REVIEW OF CURRENT OPTICAL TECHNIQUES FOR BIOMEDICAL PHYSICAL MEASUREMENTS

Douglas A. Christensen

## TABLE OF CONTENTS

## ABSTRACT

Optical radiation has unique properties which make it attractive for sensing certain physical biomedical parameters. It possesses a very short wavelength (typically in the range 0.4 to 1.0 $\mu$m), allowing accurate measurements of motion and displacements approaching this size. Interference is also minimized between the optical signal and lower frequency electrical noise usually present in the biological environment. The advent of the laser provides a means for generating narrow-frequency, spatially coherent light, and fiber optics may be used for conveniently guiding light to the sensor. Instruments capitalizing on these properties have been designed for the measurement of biological flow, force, displacement, and temperature.

Quantities such as blood flow and sperm motility can be measured by utilizing the optical Doppler shift from the moving scatterers, with high precision resulting when narrowband laser sources are used. Both air paths and fiber optic paths for in vivo measurements have been reported. For measuring small intralumen pressure changes, a thin reflecting membrane may be placed at the distal tip of a fiber-optic catheter, resulting in a varying reflected intensity as the membrane deflects in the pressure field.

Small displacements of bones and other body surfaces have been measured by interferometric techniques, either in real-time or recorded on film (holography). In a different technique, sarcomere length in excised skeletal muscle and cardiac muscle has been dynamically detected via changes in the diffraction pattern of laser light passed through the muscle.

Temperature probes using optical fibers connecting to the sensor have been reported for a variety of sensor types. The most successful employ either a liquid-crystal reflecting layer, a birefringent crystal, or a semiconductor band-edge absorber.

## I. INTRODUCTION

This paper reviews some current optical techniques for biomedical physical measurements. By this is meant the measurement of such variables as flow, displacement, pressure, and temperature. It will not cover other biomedical applications of optics such as chemical measurements (i.e., blood oximetry), imaging, endoscopy or cell recognition which, while interesting, are beyond the paper's scope. In discussing each optical technique, reliance will be made upon specific examples to illustrate the pertinent methods.

It is helpful to briefly review the unique characteristics of optical radiation which make it useful for measuring certain body parameters. First of all, it is electromagnetic radiation with a very short wavelength (0.4 to 1.0 $\mu$m) compared to radio waves. Because of this, optics offers the possibility of sensitive flow and displacement measurements on the order of a wavelength by interferometric techniques, as in holography and laser Doppler velocimetry. Second, light is easily reflected from membranes and scattered from particles, making pressure and motion monitors using thin mirrors possible and allowing the determination of RBC and sperm velocities. Finally, there is immunity to interference at the optical frequencies with the usual sorts of lower frequency background electrical noise encountered in the biological environment, such as EMG and ECG signals.

To capitalize on these advantages in biomedical instrumentation, practical devices for the generation, guidance, and detection of light are of prime importance. The laser is the best source if one needs coherent output as in laser Doppler velocimetry or holography, or if high power is required. Gaseous lasers (such as He-Ne and Argon ion) and solid-state lasers (such as Nd-Yag) have very high temporal coherence corresponding to narrow frequency spreads of only about one part per million. Continuous power outputs range from a few milliwatts to several watts.[1,1a] They are, however, expensive and somewhat bulky. Semiconductor lasers (GaAlAs, for example) are much smaller, typically being packaged in TO-5 cans, but their coherency is compromized by frequency spreads of about one part in 200. At the low end of the coherency spectrum are light emitting diodes (LEDs). Although lower in power and coherency than lasers, LEDs are very convenient and inexpensive.

An air path was traditionally the first transmission medium used. Focusing may be accomplished easily with lenses. However, air paths are susceptible to interruption in crowded settings. Therefore, for convenience in clinical situations, fibers are often the light guide of choice. Glass or plastic fibers transmit the light by total internal reflection down the core. The rapid advances being made in fiber optic communications have brought the development of low-loss optical fibers with losses of only a few dB/km. These fibers are also small (50 to 500 $\mu$m in diameter), flexible, inexpensive, and relatively durable. New fiber products are proliferating rapidly.[2]

In the area of optical detectors, vacuum photomultiplier tubes are falling into disuse now since solid state photodetectors (such as the silicon photodiode) have improved to the point of approaching the stability and low noise of tubes. They have the added advantages that they are inexpensive and quite small.[3] Figure 1 illustrates some elements which might be used in a current optical instrument, showing a loop of optical fiber, a light emitting diode, and a silicon photodetector. The fibers cost only a few cents, the LED is $1.95, and the photodetector is about $12.50.

To best illustrate the current art of optical biomedical measurement techniques, some specific examples will next be described.

## II. PRESSURE: FIBER OPTIC CATHETER-TIP TRANSDUCER

The fiber optic catheter-tip pressure transducer was originally designed by Morikawa et al.,[4,4a,4b] for cardiovascular measurement. A detailed sketch of the sensor is given in Figure 2. The principle of operation is fairly straightforward. A highly metallized 20-$\mu$m thin glass membrane is positioned at the distal tip of a fiber optic catheter with a small air space between the ends of the fibers and the reflecting membrane. The catheter, with an outside diameter of 7 French, contains several fibers for transmitting light and several fibers acting as receivers. As the membrane deflects inwardly due to the imposed pressure, it will reflect a varied amount of light back into the fibers. An experimental study of the returned light intensity as a function of spacing between membrane and fibers shows an almost linear relationship in the range of 50 to 100 $\mu$m spacing. If the sensor were biased at zero pressure to be near the center of this range, the optical output would be reasonably linear with displacement. In turn, a mathematical treatment of the displacement of thin membranes with pressure shows that in the center portion of the membrane the displacement is proportional to pressure for motions of less than half the membrane thickness. Thus, optical output should be closely linear with pressure. A prototype model has proven this to be true over the pressure range of 0 to 150 mmHg. Sensitivity can be increased by using thinner membranes, but rupture pressure is reduced also.

The dynamic response of the sensor may be compared to a standard commercially available transducer by placing both in the same cavity and imposing a step pressure change on the cavity. Such a test performed with a prototype fiber optic sensor showed at least as high a frequency response as the commercial sensor.

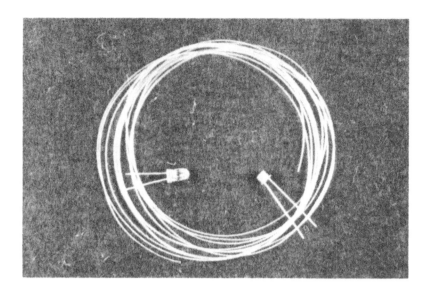

FIGURE 1.   Typical components which might be used in a present-day electrooptic instrument, showing a loop of plastic fiber, an inexpensive light-emitting diode, and a silicon photodetector.

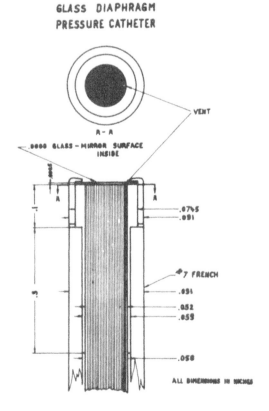

FIGURE 2.   A schematic diagram showing the essential elements of the catheter-tip pressure transducer.

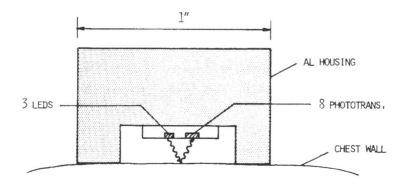

FIGURE 3. A simplified diagram of the arrangement of optical components employed in the electrooptic cardiac sound sensor.

## III. MOTION: ELECTROOPTICAL CARDIAC SOUND SENSOR

The analysis of the frequency content of cardiac sounds is an important clinical diagnostic tool.[5] Conventional techniques use a wideband microphone taped to the patient's chest, but such an arrangement is awkward since the microphone is usually heavy. An optical reflection scheme, similar to the preceding reflecting membrane idea, has been developed by Yeung et al.[6] A simplified diagram is given in Figure 3. Here, light from three simultaneously excited LEDs is reflected from the skin back to an array of eight phototransistors connected in parallel. The multiple number of transmitters and receivers is used to increase the signal-to-noise ratio and decrease the sensitivity to specific skin reflection characteristics. As the chest wall vibrates, a time-varying amount of light will be detected, obeying the inverse square law with distance. In an experimental model, the minimum detectable motion (determined by electrical sensitivity) was found to be 0.225 mm.

Advantages of this technique include a wide bandwidth (about 3 to 9 kHz), economy, and potential small size and weight. Expected improvements in the signal-to-noise ratio as the optical and electronic components are improved may allow an even smaller array size and package weight.

## IV. DISPLACEMENT: OPTICAL HOLOGRAPHY

Holography using light is a more sophisticated way of measuring displacement.[7] A diagram showing the essential arrangement needed for both holography and laser Doppler velocimetry is given in Figure 4A. The technique is based upon the interference between two beams. One is generally a uniform reference beam, and the other is a beam reflected from the surface of the object under investigation. The wavefronts of the reflected beam ("object" beam) take on the shape of whatever object is being studied. In regions where they overlap, the beams interfere with one another in such a way that where the wavefronts are in phase they constructively interfere, causing a bright fringe. Where the wavefronts are 180° out of phase, they destructively interfere, causing a dark fringe. If the object is nonmoving, these fringe positions are stationary in space even though the radiation is moving through space at the speed of light. Therefore, the positions of the fringes can be recorded on high-resolution film, producing what is known as a hologram. In order to achieve sharp fringes, the illumination must possess well-defined phase fronts, i.e., it must have a high degree of coherence.[8] Thus, a laser is almost universally used as the source for holography.

INTERFEROMETRIC   BASIS   FOR   HOLOGRAPHY

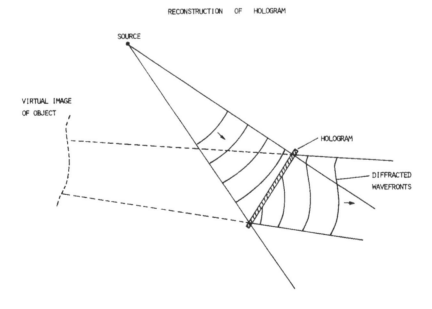

FIGURE 4 (A).   A simplified sketch of the interferometric basis for holography, showing how the beam reflected from the object will interfere with the reference beam to produce dark and bright fringes, which may be recorded on film (a hologram).

(B).   Reconstruction of the original object wavefronts can be achieved by reilluminating the developed hologram with a replica of the reference wave.

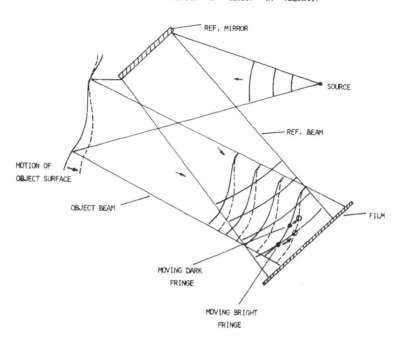

MOTION OF OBJECT IN HOLOGRAM

(C). If motion of the object occurs, the dark and bright fringes will also move. This fringe movement may be measured in three different ways, as discussed in the text.

To reconstruct the object, the hologram is reilluminated after the film has been developed, using the setup shown in Figure 4B. The reconstruction beam passes through the hologram, and the fringes which have been recorded on the grains of the film diffract the laser light as a diffraction grating into wavefronts at side angles. These diffracted wavefronts will reproduce exactly the wavefronts that were reflected from the original object, thus producing a virtual image of the object. Since these wavefronts are exact, the virtual image will be three dimensional and have all the characteristics of the original object. Thus, positional information is recorded for observation and analysis. However, of greater interest for this paper is the measurement of the *movement* of the object or the displacement of the object's surface.

Figure 4C shows what happens when there is motion of the object's surface. As before, interference fringes are generated at the position of the film, but movement of the object will cause the reflected wavefronts to interfere with the reference beam at different positions than before. Thus, the dark and bright fringes will move in space if the object is moving. To take advantage of this phenomenon for motion measurement, the recording technique may take one of three different forms, as described next.

The first way is to leave the film in place and let the moving fringes blur across the film. Every place on the film that corresponds to a stationary spot on the object will record nonmoving fringes. On the other hand, every place that corresponds to movement of the object will show fringes that have moved and are blurred during the time exposure of the film. When the object is reproduced upon reconstruction of the hologram, all moving portions of the object will look blurred and somewhat darker, and all stationary spots will appear to be in focus. This gives a semiqualitative way of looking at the movement of biological systems. For example, Feleppa has used this technique to observe regions of movement in slime mold where the regions of motion-induced darkness are apparent in the reconstructed image.[9]

A way of quantitatively measuring the movement of the object is to expose the hologram at two particular times, $t_1$ and $t_2$, when the object is in two different positions. This is called double-pulse holography. The hologram will record the superposition of the two fringe patterns, and upon reconstruction, the two virtual images (one recorded at $t_1$ and one recorded at $t_2$) will themselves interfere if movement occurred, causing another interference pattern whose fringes overlie the average virtual image. Any small displacement on the order of one half of a wavelength or more which takes place between times $t_1$ and $t_2$ will cause fringes to appear in the image, and the spacing and position of the fringes can be quantitatively related to surface movement. Such a double-pulse technique has been reported to observe very small displacements (about 0.5 µm) in teeth and parodontally anchored prosthodontic appliances during normal force production in mastication.[10] One subject had a tooth stressed during double-pulse holography, and the resulting fringe pattern was sensitive enough to display a distal displacement of only 1.0 µm. Conceivably, this technique could be used to measure small displacements in any exposed bony surface.

The third method is actually a configuration known as laser Doppler velocimetry and is detailed in Figure 5. An opaque screen at the position of the reference beam/object beam interference pattern blocks all of the fringes except for that light which passes through a pinhole whose diameter is small compared to a fringe width. The shifting fringe pattern will project alternately dark and bright fringes through the pinhole, and the photodetector will detect an alternating intensity. The faster the object moves, the higher the frequency output of the photodetector. The object may be replaced by a collection of scatterers, such as RBCs or moving sperm. In this configuration, the device is known as a homodyne laser Doppler velocimeter and has found many applications in biological motion measurements, as discussed next.

## V. VELOCITY: LASER DOPPLER FLOWMETERS

The first laser Doppler instruments used an air path, such as the one studied by Morikawa et al.[11] The goal of the project was to map the pulsatile flow in an in vitro model of the cardiovascular system. Figure 6 shows the system with the laser source focused by a lens into a section of glass tubing where there was pulsatile flow of water seeded with artificial scatterers. For Doppler interference, there must be another beam, so a beam splitter was used to provide a reference beam. This was then recombined with the scattered beam to yield interference through an aperture on the face of the photomultiplier tube. The output spectrum from the photomultiplier tube was then studied, since from the Doppler shift principle the frequency difference between the reference beam and the scattered beam is proportional to the velocity of the scatterers as they pass through the flow section. One advantage of an air path is the ease with which a lens can be used to focus the beam to a very small volume, about 30 µm in diameter, thereby obtaining localized flow in the region of artificially induced disturbances. For example, in one case a catheter was placed in the flow, and disturbances around the tip of the catheter were studied. Similarly, flow information around valves, bifurcations, and restrictions may be studied. The small sensing region allows detailed flow profiles across the tubing diameter to be mapped, a measurement which is very difficult with other techniques like electromagnetic or ultrasonic flowmeters. A disadvantage of this method, however, is the requirement for transparent tubing. A later modification of this laser velocimeter utilized a frequency-translated reference beam to resolve the ambiguity over the direction of velocity that is inherent in the conven-

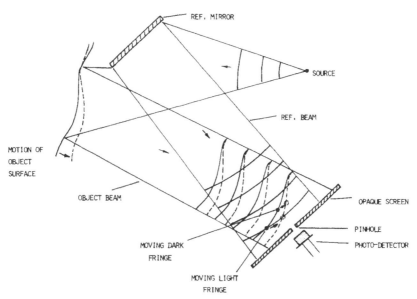

FIGURE 5. The interferometric basis for laser Doppler velocimetry measurements, where a photodetector measures the time variation in intensity passing through the pinhole as the dark and bright fringes move past it. Compare the similarities between this diagram and Figure 4 for holography.

## OPTICAL SYSTEM

### FOR THE LASER DOPPLER VELOCIMETER

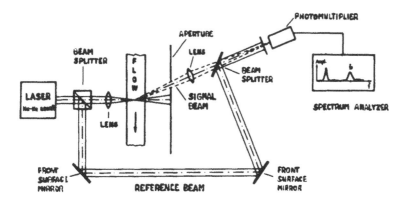

FIGURE 6. A schematic of the components used in an air-path laser Doppler velocimeter measurement of flow in a cardiovascular model.

tional technique. This bidirectional flowmeter allows determination of the flow direction, important in heart-valve simulation studies.[12]

Air paths are not very convenient in actual clinical instruments because obstructions may occur in the air paths and because of the need for a rigid arrangement of focusing lenses. A much better technique is to use flexible fiber optics as the transmission path. Figure 7 is a schematic diagram of a device for measuring surface blood flow in the

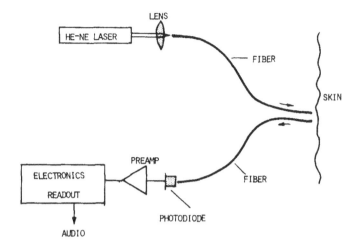

CUTANEOUS LASER DOPPLER BLOOD FLOW

FIGURE 7.    The essential elements of a cutaneous laser Doppler
blood flowmeter using a pair of optical fibers.

skin developed by Holloway and Watkins.[13] The light from a He-Ne laser is coupled
into one of a pair of optical fibers and is transmitted to the surface of the skin. Part
of the incident light is reflected off the skin's surface back into the receiving fiber,
and this becomes the reference beam. The skin essentially represents a nonmoving
object giving a nonshifted reference signal to the receiving fiber. In addition, part of
the light is transmitted through the cutaneous layers where it is reflected off moving
erythrocytes in the microcirculation and is also collected by the receiving fiber. The
authors estimate that this reflection takes place at a depth of from 1.0 to 1.5 mm into
the skin. At the photodiode, the two beams in the fiber interfere and the Doppler-shift
frequency is amplified for appropriate readout. (In this particular instrument, it is also
possible to listen to the Doppler frequency on a set of earphones.) Experimental results
show that this technique for measurement of the velocity of the microcirculation cor-
relates well with the xenon-clearance method.

The configuration just described is rather unique in that it is one of the few situations
where optics can actually penetrate a short distance into the body. Another location
where optics can penetrate without invasiveness is into the eye's retina for the measurement of retinal circulation.

Another example of a fiber optic velocimetry system is that of Tanaka and Benedek,[14,14a] described in detail in another section of this book. It consists of a single fiber
fiber introduced into a vein. A laser beam is focused by a lens onto the end of the
fiber for transmission to the vein. Part of the light is reflected backwards in the optical
by Fresnel reflection from the fiber's distal end and becomes the unshifted reference
beam. The remaining light is transmitted into the blood where it is reflected from the
moving erythrocytes and recollected by the fiber. A rather clever configuration is used
at the fiber tip in order to make the incident light vector nonparallel to the axis of the
fiber. It consists of an oblique cut at the end of the fiber from which light is reflected
by total internal reflection out the sides of the fiber.

The reference and scattered beams in the fiber are then sent to the face of a photomultiplier tube by a beam-splitter arrangement. A correlation technique yields the
Doppler frequency and, in turn, the cell velocities.

Laser velocimetry is not limited solely to blood flow. Shimizu and Matsumoto measured sperm velocities using a technique similar to those described above.[15] Light from a helium-neon laser is passed through a polarizer and neutral density filter, then focused by a lens into the sample cell where the sperm are placed. The light scattered at an angle $\theta$ is then passed through an analyzer and two slits to a photomultiplier tube whose output is analyzed by a minicomputer using correlation techniques. Again, good agreement is found between this method and more cumbersome and time-consuming techniques such as direct microscopic observation.

## VI. FLOW: URINE DROP FLOWMETER

Another optical flowmeter which is not a laser Doppler flowme er, but has important clinical applications, is shown in Figure 8. Developed by Ritter et al., the unit measures the flow of urine and is fairly simple.[16,16a] It utilizes the drops which develop naturally in the voided stream to interrupt a stream of light. An earlier version used fiber optic sheets for transmission and reception of the light, but the present instrument forms the light from a high-intensity bulb into a sheet by first colimating it with a cylindrical lens, then restricting its vertical extent with a slit. After passage across the voiding space, the light is passed through a second slit to a long rectangular photodiode detector. An analysis of the light detected will yield volume vs. time data, from which information relating to possible urethral obstructions may be obtained.

## VII. DISPLACEMENT: SARCOMERE LENGTH

Another example of measuring displacement that might not be as well-known as holography is an unique use of optics in measuring muscle sarcomere length by diffraction. This measurement is based upon the fact that both skeletal and cardiac muscle have striations when viewed under a light microscope which are caused by phase differences between the A and I bands, the overlapping and nonoverlapping regions of the filaments in the muscle. If coherent light from a laser is passed through a few fibers of such muscle, some of it will be diffracted into side orders similar to those from a conventional diffraction grating. The sine of the angle $\theta_n$ between the $n^{th}$ order and the original propagation direction will be inversely proportional to the spacing, d, between the striations, according to the formula

$$\mathrm{Sin}\ \theta = n\lambda/d$$

where $\lambda$ is the wavelength of the incident light, and $n = 0, \pm 1, \pm 2, \ldots$. Figure 9 is a schematic diagram showing a laser beam being passed through a muscle sample. The lens collects both the undiffracted and diffracted light and focuses it on the observation screen where the central undiffracted order ($n = 0$), the first ($n = \pm 1$), and the second ($n = \pm 2$) orders can be seen.

Investigators studying muscle dynamics are currently interested in how striation spacing reacts as a function of time after the excitation of the muscle. Pollack et al. have assembled an experimental setup where a muscle sample is placed in a microscope stage, laser light is passed upwards through the bottom of the stage and the sample, the orders are collected by the microscope objective lens, and the diffraction pattern is focused onto a TV camera.[17,17a] A video tape recorder records the pattern which is later used for analysis of the time history of the sarcomere length upon excitation of the muscle. Using the diffraction formula given above, a measurement of the angle to the peak of the first order will yield the average sarcomere length of the sample. There is also some indication that the angular width of the orders reflects the distribution of sarcomere lengths in the illuminated segment of the muscle.

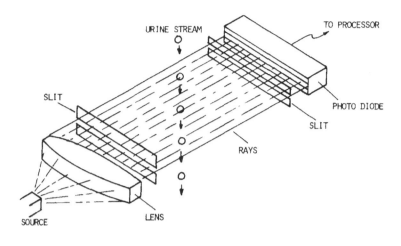

URINE   DROP   FLOWMETER

FIGURE 8.   A design for a urine flowmeter using a cylindrical lens and slits to produce a thin sheet of light for sensing drops.

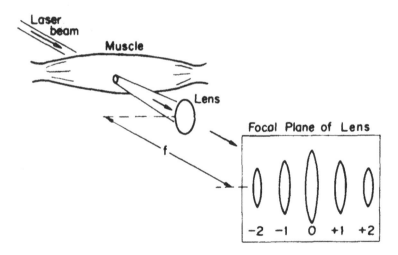

FIGURE 9.   The use of laser diffraction from the phase striations in cardiac muscle to provide information about the dynamics of sarcomere length changes in contracting muscle. The angles of the diffracted side orders are inversely related to sarcomere length.

## VIII. TEMPERATURE: FIBER OPTIC TEMPERATURE PROBES

Although temperature measuring devices with metallic components, such as thermistors and thermocouples, are plentiful, economical, and accurate, there is the need in certain applications for sensors which avoid any electrically conducting parts. These applications include microwave and RF experiments (for example, biohazard studies and possible cancer therapy involving microwave-produced hyperthermia) where the metallic components will perturb the electromagnetic fields leading to false temperature readings and clinical situations (such as catheter measured thermodilution deter-

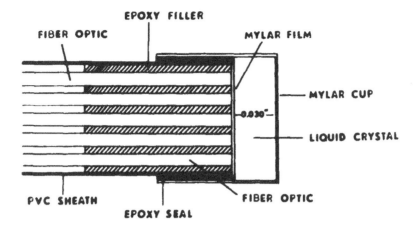

FIGURE 10. A diagram of the tip region of the liquid crystal fiber optic temperature probe, showing the thin layer of reflecting liquid crystal material.

mination of cardiac output) where concern about electrical shock dictates strict safety precautions. In these cases, a nonconducting method for temperature measurement is needed, leading to the development of fiber optic temperature sensors.

Several different types of sensors have been reported, notably the liquid crystal optical fiber probe,[18] the birefringent crystal temperature sensor,[19] the etalon optical probe,[20] and the semiconductor band-edge absorption sensor.[21] All of these sensors are meant to be used with an optic fiber bundle as the transmitting means. Here, we will discuss only the first and last types of sensors.

The liquid crystal probe was one of the earliest systems developed by Johnson, Rozzell, and others.[18,18a] It is based upon the change in color reflectivity demonstrated by a layer of cholesteric liquid crystal material as the temperature of the layer varies. In fabrication, a thin coating of liquid crystal is sandwiched between two nested glass bulbs at the tip of a plastic multifiber bundle, and red ($\lambda = 0.670\ \mu m$) light from an LED is transmitted down a few of the fibers, reflected from the tip into the remaining fibers, and detected by a photodiode. Figure 10 shows a schematic drawing of the tip region. As the temperature (and thus color) of the reflecting layer varies, the amount of light detected varies. By altering the composition of the components in the liquid crystal and adjusting the transmitted wavelength, optimum sensitivity can be obtained for any chosen range of temperatures within the capabilities of the mixture. A useful range for biological work includes 32 to 45°C. Within that range, a precision of ±0.1°C has been achieved.

However, improvements in the liquid crystal probe are needed to overcome some drift problems (apparently caused mainly by instabilities in the chemical layer itself) and to fabricate the tip in smaller sizes (presently limited to about 1.5 mm in diameter).

The semiconductor band-edge probe is based upon band-edge absorption in a bulk semiconductor sample, which may be explained by reference to Figure 11. Indicated here is the energy gap diagram for a direct-gap semiconductor crystal showing the almost-filled valence band (at normal temperatures), a forbidden energy zone, and the almost-empty conduction band. When light is passed through a sample of this semiconductor, it will be absorbed by an amount dependent upon the degree that its photon energy exceeds that needed to excite valence band electrons into the conduction band. The difference in wavelength between heavy absorption (at shorter wavelengths) and low absorption (at longer wavelengths) is very slight. In addition, the energy of the band gap changes as a function of temperature. Therefore, as narrow-bandwidth light

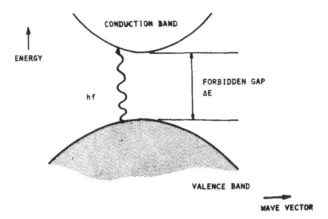

FIGURE 11.    An energy gap diagram for a semiconductor. Photons will be absorbed if their energy is sufficient to excite valence band electrons into the conduction band.

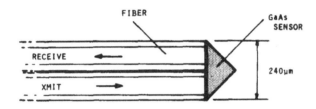

FIGURE 12.    A diagram of the tip region of a semiconductor temperature probe using optical fibers.

with a center wavelength positioned on the absorption slope is transmitted down the optical fibers and through the sensor to be received by the receiving fibers for detection and display, the intensity detected is related to the temperature of the tip.

The principle was used by Christensen[21] for the development of a small-tipped probe whose sensor region is diagrammed in Figure 12. The simplicity of the design leads to a small tip diameter, currently about 0.300 mm, giving minimum tissue trauma when used in animal or human experiments. The sensor material is low conductivity GaAs, and a 0.905 $\mu$m wavelength (near IR) LED is employed. The range of the instrument is from 20 to 50°C, with an accuracy of 0.1°C and a resolution of 0.01°C. Long-term stability of the unit is dependent only upon the electrooptics of the system, since the semiconductor sensor itself is inherently stable. The pressure dependence of the band gap is very slight, since a pressure change of 100 mmHg corresponds to less than a 0.01°C equivalent change in the gap. Thus, the sensor is essentially sensitive only to temperature.

The examples discussed in this paper of optics at work measuring various physical biomedical parameters have been chosen to demonstrate the variety and creativity of this mode of instrumentation. Future years will undoubtedly contain even more developments using optics in biomedical measurement systems.

# REFERENCES

1. O'Shea, D. C., Callen, W. R., and Rhodes, W. T., *Introduction to Lasers and Their Applications*, Addison-Wesley, Reading, Mass., 1977.

1a. Wolbarst, M. L., Ed., *Laser Applications in Medicine and Biology*, Vol. 1, Plenum Press, New York, 1971.

2. Barnoski, M. K., Ed., *Fundamentals of Optical Fiber Communications*, Academic Press, New York, 1976.

3. Williams, C. S. and Becklund, O. A., *Optics: A Short Course for Engineers and Scientists*, Wiley-Interscience, New York, 1972.

4. Morikawa, S., Ofstad, J., Johnson, C., and Sandwith, C., Fiberoptic Catheter-Tip Pressure Transducers, 8th ICMBE, Chicago, July 25, 1969.

4a. Lundstrom, L. H., Miniaturized pressure transducer intended for intravascular use, *IEEE Trans. Biomed. Eng.*, July 1970, 207.

4b. Zubareva, T. V., Urmakher, K. S., and Shapiro, E. I., Fiber-optic tonometer for measuring intraocular pressure, *Med. Tekh.*, 2, 21, 1973.

5. Adolph, R. J., Stephens, J. F., and Tanaka, K., The clinical value of frequency analysis of the first heart sound in myocardial infarction, *Circulation*, 41, 1003, 1970.

6. Yeung, S. K., Yee, S., and Holloway, A., An electrooptical sensor for cardiac sound and vibrations, *IEEE Trans. Biomed. Eng.*, January 1977, 73.

7. Gabor, D. and Stroke, G. W., Holography and its applications, *Endeavor*, 27(103), 40, 1969.

8. Goodman, J. W., *Introduction to Fourier Optics*, McGraw-Hill, San Francisco, 1968.

9. Feleppa, E. J., Holography and medicine, *IEEE Trans. Biomed. Eng.*, May 1972, 194.

10. Wedendal, P. R. and Bjelkhagen, H. I., Dynamics of human teeth in function by means of double pulsed holography; an experimental investigation, *Appl. Opt.*, 13(11), 2481, 1974.

11. Morikawa, S., Lanz, O., and Johnson, C. C., Laser Doppler measurements of localized pulsatile fluid velocity, *IEEE Trans. Biomed. Eng.*, November, 1971, 416.

12. Lanz, O., Johnson, C. C., and Morikawa, S., Directional laser Doppler velocimeter, *Appl. Opt.*, 10, 884, 1971.

13. Holloway, G. A. and Watkins, D. W., Laser Doppler measurement of cutaneous blood flow, *J. Invest. Dermatol.*, 69(3), 306, 1977.

14. Tanaka, T. and Benedek, G. B., Measurement of the velocity of blood flow (in vivo) using a fiber optic catheter and optical mixing spectroscopy, *Appl. Opt.*, 14(1), 189, 1975.

14a. Tanaka, T., Riva, C., and Ben-Sira, I., Blood velocity measurements in human retinal vessels, *Science*, 186, 830, 1974.

15. Shimizu, H. and Matsumoto, G., Light scattering study on motile spermatozoa, *IEEE Trans. Biomed. Eng.*, March 1977, 153.

16. Ritter, R. C., Zinner, N. R., and Sterling, A. M., The urinary drop spectrometer — an electrooptical instrument for urological analysis based on the external urine stream, *IEEE Trans. Biomed. Eng.*, May 1976, 266.

16a. Zinner, N. R. and Harding, D. C., Velocity of the urinary stream, its significance and a method for its measurement, in *Hydrodynamics of Micturition*, Hinman, F., Ed., Charles C Thomas, Springfield, Ill., 1971, chap. 16.

17. Krueger, J. W. and Pollack, G. H., Myocardial sarcomere dynamics during isometric contraction, *J. Physiol.*, 251, 627, 1975.

17a. Krueger, J. W., Christensen, D. A., and Pollack, G. H., Myocardial sarcomere length measurements by spatial fourier spectral analysis, *Proceedings of the 26th Annual Conference on Engineering in Medicine and Biology*, Alliance for Engineering in Medicine and Biology, 1973, 97.

18. Johnson, C. C., Durney, C. H., Lords, J. L., Rozzell, T. C., and Livingston, G. K., Fiberoptic liquid crystal probe for absorbed radio-frequency power and temperature measurement in tissue during irradiation, *Ann. N.Y. Acad. Sci.*, 247, 527, 1975.

18a. Rozzell, T. C., Johnson, C. C., Durney, C. H., Lords, J. L., and Olsen, R. G., A nonperturbing temperature sensor for measurements in electromagnetic fields, *J. Microwave Power*, 9, 241, 1974.

19. Cetas, T. C., Hefner, D., Snedaker, C., and Swindell, W., Further developments of the birefringent crystal optical thermometer, *1976 U.S. National Committee/ International Union of Radio Science Meeting*, National Academy of Science, Washington, D.C., 1977, 11.

20. Christensen, D. A., Temperature measurement using optical etalons, *1974 Annual Meeting of the Optical Society of America*, Optical Society of America, New York, 1975.

21. Christensen, D. A., A new nonperturbing temperature probe using semiconductor band edge shift, *J. Bioeng.*, 1, 541, 1977.

*Applications of Physical Sensors in Life Sciences*

Chapter 7

# THE APPLICATION OF PHYSICAL SENSORS TO STUDIES OF THE CARDIOVASCULAR SYSTEM

## Ernest P. McCutcheon

## TABLE OF CONTENTS

# I. INTRODUCTION

The purpose of this presentation is to initiate and stimulate discussion of the physical sensors used to obtain significant information on cardiovascular function. It is not an encyclopedic review of all possible applications to this area, since virtually every type or class of physical sensor has been used or has the potential for use in cardiovascular studies. It is intended to provide a representative example of the broad applications of physical sensors in a specific specialty area. My viewpoint is that of the user community rather than the instrumentation design engineer. Examples have been chosen to illustrate particular classes of problems and to demonstrate uncertainties created by shortcomings of existing sensing techniques. Emphasis is on the more pragmatic and realizable rather than the theoretical and optimal. Human applications have received considerable emphasis in earlier chapters, and for that reason as well as my personal research orientation, applications in animal studies will be stressed.

In every case, certain background considerations always apply even if they are not mentioned explicitly. Factors to be kept in mind affecting the choice of sensors and implementation of sensor systems include (1) the degree to which the characteristics of the variable are static or dynamic, (2) whether the structure or organ sampled is isolated and perfused in a bath, or within the intact organism, (3) the extent to which the required approach is noninvasive or invasive and implantable, (4) the length of time required for the measurement, ranging from brief and acute to long-term and chronic, and (5) whether the sensor output is accessed through a directly wired connection or transmitted through a telemetry link. Such considerations always impact the measurement process and illustrate the broad range of factors affecting applications of physical sensors in analysis of cardiovascular function.

The selected examples of cardiovascular measurements have been grouped into three primary areas. First is force and its spatial or structural distribution as tension (force/ length), pressure and stress (force/area), and the proportional displacement of strain ($\Delta$ length/initial length). Second is the determination of volumes and the movement of fluids and structures, an aspect also including sound. The third area includes the biopotentials and electrical activity, and the thermal processes underlying all cardiovascular function, about which very little will be said in this paper since the emphasis is on sampling of physical-mechanical events.

# II. FORCE, DIMENSIONS, AND PRESSURE

## A. Force and Dimensions: Isolated Tissue

Let us first consider an example illustrating the class of an isolated tissue specimen in a comparatively highly controlled environment, a configuration often taken for granted (Figure 1).[1] This approach is particularly suitable for the determination of force, tension, and stress/strain properties. This is a very significant and fundamental technique for obtaining information about cardiovascular function, and perfused segments of vessels or heart muscle are frequently studied in this manner.[1-7] The configuration illustrated in Figure 1 is for the study of papillary muscle.[1] One of the notable features is its horizontal orientation, technically somewhat easier to manage than a vertical system. The force transducer is incorporated in the short, rigid lever arm. A muscle lever with clamp (not shown) is attached to the transducer mounting arm to hold one end of the specimen. The other clamp (shown) holds the opposite end of the tissue sample, and a micrometer adjusts the overall specimen length. The pair of clamps holds the sample in the isometric state so that it does not shorten with contraction. Thus, the transducing system must have very low compliance. Frequently, such

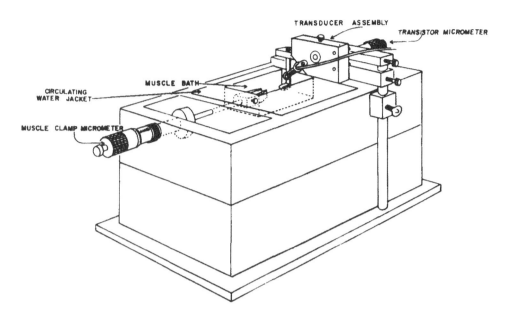

FIGURE 1. Temperature controlled bath for study of isolated tissue specimens. The force transducer (Pitran transistor) is mounted in the short, rigid lever arm. A muscle lever with clamp (not shown) secures one end of the specimen and is placed so that its projecting bead activates the transducer. The transistor micrometer adjusts the muscle lever-transistor diaphragm contact. The clamp shown at the opposite end of the bath holds the other end of the specimen, and a clamp micrometer adjusts overall specimen length. The transducer output was linear from 0 to 1 kg and frequency response was mechanically limited to the range of DC to 1 kHz. (From Jacobs, H. K., McConnell, D. P., Rowley, B. A., and South, F. W., *J. Appl. Physiol.*, 35, 1973, 436. By permission.)

systems have used a deflecting beam with an attached strain gauge for measurement of the contraction-generated force. A special feature of the device of Figure 1 is the linear semiconductor element (Pitran transistor) incorporated in the fixed transducer mounting arm to provide a highly sensitive detector. A very stable system with a capacitive transducer has also been described.[2] One of the nicest things about this bath preparation is that the desired dimensions such as length, circumference, and cross-sectional areas can be determined easily by a variety of techniques, from rulers and calipers[3-5] to optical, noncontacting approaches such as those described in the Workshop paper by Christensen. Results of such experiments have been valuable in documenting the characteristic exponential stress-strain relationship of cardiovascular structures and the linearity of the elastic modulus vs. stress. Interestingly, a papillary muscle in a configuration of this general type was thought initially to contract homogeneously, but on closer examination, the external fibers are observed to shorten while those in the central core lengthen, so that even isolated samples can be very complicated systems.[6,7] Testing of vessel pieces under similar conditions has quantified differences between veins and arteries, and the increased passive arterial stiffness and other changes with age and atherosclerosis.[8,9]

## B. Force and Dimension: Intact Ventricle

Similar determinations in other geometries and in intact systems are considerably more difficult because mechanical devices readily produce tissue loading and varying degrees of error. The thick ventricular walls should be least susceptible to such effects, and a variety of devices have been applied. Figure 2 shows the heroic lengths to which physiologists must often go in order to get data. This example is the classic resistive

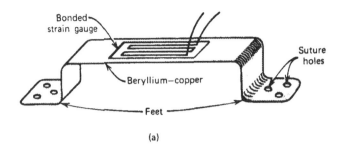

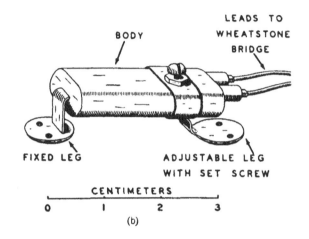

FIGURE 2.   Strain gauge arch for measurement of muscle force: (a) construction principles, (b) encapsulated gauge with adjustment of muscle length by altering the distance between the legs. Frequency response was 45 Hz. Minimizing temperature dependence requires balanced strain elements. The force magnitude depends on the suture depth and tightness. (From Cotten, M. deV. and Bay, E., *Am. J. Physiol.*, 187, 122, 1956, 123; and **Cobbold, R. S. C.**, *Transducers for Biomedical Measurements*, John Wiley & Sons, New York, 1974, 179. By permission.)

strain gauge arch sewn into the muscle.[10,11] The strain gauge is used in the bending mode. Adjustment of the legs stretches the muscle and provides a quasi-isometric configuration analogous to the bath situation.

A successor device avoided the problems of variable suture tension, but was considerably more fearsome looking, spearing the depth of the myocardium and directing the load to deflection plates on which strain gauges were bonded (Figure 3).[12] The contracting muscle is attempting to pull the pins together without any real movement so the output is a close approximation to the isometric condition (Figure 3A).[12] In related designs, the pins are positioned close together, unloading the short muscle segment between them. (Figure 3B).[12] The device is essentially in series with the contracting muscle (auxotonic condition), sensing its tendency to pull apart and measuring the circumferential force.[12,13] A significant result with such approaches was the demonstration of the time-dependent nature of myocardial wall force development, with a more rapid increase and decline in tension than pressure during the cardiac cycle. Assessment of the significant dynamic stress/strain properties requires measurement of wall thickness, which also varies with the changing stresses of the cardiac cycle. One early approach is shown in Figure 3C.[14] A thin, arched brass shim was mounted on a spring-loaded circular plate sliding on a steel shaft. The shaft was thrust through the ventric-

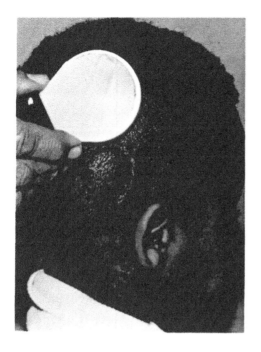

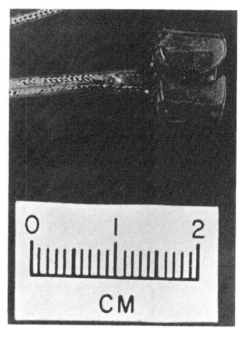

FIGURE 10 (Chapter 5). ICP device and patient after implant surgery.

FIGURE 15 (Chapter 7). A durable, reusable prototype ultrasonic Doppler transducer for application to small vessels such as a coronary artery. The transducing elements are cast within a durable, hyporeactive epoxy resin.

FIGURE 1 (Chapter 8). Close-up of integrated circuit force transducer assembly. At upper left is the active element with integral suture loops. The attachment of platinum-iridium wires to pads on the transducer can be clearly seen. At the lower right is the silicone rubber block with a perforation for ease in suturing.

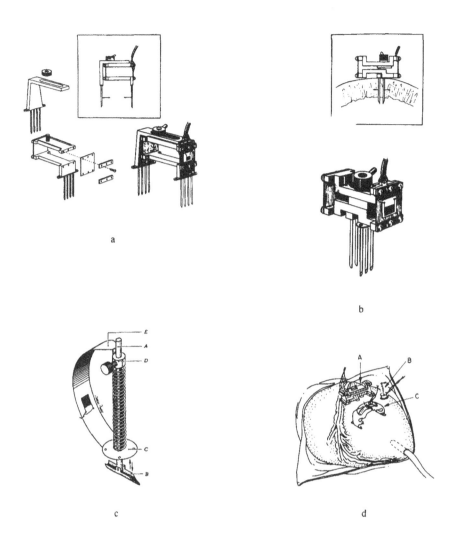

FIGURE 3. Cardiac force transducers. The muscle was coupled to the deflecting plates by pins inserted through the ventricular wall. The system was linear, and frequency response was flat to 30 Hz. (a) Isometric design: pin separation stretches the intervening muscle and limits shortening. (b) Auxotonic type: moving the pins together compresses the intervening segment, limiting its force producing capability and placing the transducer in series with the remainder of the muscle for estimation of the circumferential force. The arrows indicate the direction of the muscle force. (From Feigl, E. O., Simon, G. A., and Fry, D. L., *J. Appl. Physiol.*, 23, 59, 1967, By permission.) (c) Device for measurement of myocardial wall thickness. Etched foil strain gauge elements are deflected by movement of a brass shim (E) attached to the spring-loaded circular plate (C) sliding on a slender steel shaft (A). The shaft was thrust through the ventricular wall and held within the endocardium by the toggle end-piece (B). The circular plate was sewn to the endocardium and the adjustable spring-tension (D) held the toggle against the inner ventricular surface. (From Feigl, E. O. and Fry, D. L., *Circ. Res.*, 14, 541, 1964. By permission.) (d) The sampling limitations of relatively large strain gauge devices are evident in this diagram illustrating simultaneous measurement of left ventricular wall force, thickness, epicardial arc length, and pressure. (From McHale, P. A. and Greenfield, J. C., Jr., *Circ. Res.*, 33, 1973, 305. By permission.)

ular wall and held within the ventricle by a toggle end-piece. The circular plate was sewn to the epicardium, and the spring held the toggle against the ventricular endocardial surface. The motions of the shaft are transferred to the shim for detection with the strain element. A related caliper type for surface attachment has been used for dimension measurement on the heart and aorta.[15,16] Somewhat less bulky electromagnetically coupled coils constrained to axial motion by a rigid connecting shaft and other alternative approaches for ventricular wall thickness have been applied in open-chest dogs.[16-18]

Three of the mechanical devices described were combined to measure wall force, wall thickness, and external arc length simultaneously in the left ventricle of open-chest dogs.[16] As shown in Figure 3D, their bulk and the trauma of application prevent sampling from an identical point and obviously prohibit a significant number of samples from multiple points.

Besides creating changes in volume and wall thickness, contraction of the cardiac muscles requires internal fiber rearrangement. The changing fiber orientation and thickness set up shear forces within the wall — here we reach a really desperate situation as far as sensors are concerned (Figure 4). For this case, shear in the contracting muscle was estimated from the moment applied to a spear thrust into the myocardium.[19] The loads are transmitted to a deflecting beam and, again, sensed by the ubiquitous strain gauge.

All such devices require extremely careful and complete calibration, both *in situ* and on the bench.[12,14,16] Varying degrees of tissue loading affect output waveforms and amplitudes. The inertial and elastic loads of these mechanical devices are relatively high, and they are unsuitable for applications on very compliant structures such as veins or the atria.

To obtain continuous, dynamic dimension measurement with a closer mechanical match to tissue, highly compliant strain gauges have been applied. The classical example is the resistance device consisting of silicone rubber tubing filled with mercury. Change in length of the tubing alters the conformation of the mercury column and the resulting change in electrical resistance is detected.[11,20] These devices are relatively inexpensive and simple to make, and attachment produces minimal interference with function. However, they are not very durable. As in the case of the bonded strain gauge transducers, the precise mechanical effects on the tissue are difficult to assess because of uncertainties in contact conditions and dynamic gauge impedance. Nevertheless, even for the mercury-in-rubber transducer, such potential sources of distortion appear negligible only for the ventricles and large arteries.[20]

The examples given above are falling into disuse. They are almost exclusively restricted to acute investigations in anesthetized animals, and alternative techniques are becoming more generally available. However, I know of no other worthwhile device developed for estimation of shear. For thickness determination, ultrasonic approaches are being accepted increasingly. In addition to high frequency response, the small ultrasonic sensing elements placed directly on the tissue have comparatively low mechanical loading or impedance and function under chronic or long-term conditions. Moreover, ultrasound can be used transcutaneously as described by Donald Baker. Ultrasonic dimension measurements are based on the fact that ultrasonic waves travel through materials at a measurable velocity (c), and time (t) can be converted to distance (d).[11,21] The equation is

$$d = ct.$$

The ultrasonic energy, generated by electrical excitation of disks made from piezoelectric materials, is propagated essentially in a straight line as a highly confined pencil-type beam. Practical systems are available based on either the transmission or reflection attributes of the ultrasonic energy.[11,21,22] Since the velocity of ultrasound in body tissue is approximately constant, the transit time between the transducers is a function of their separation. Devices based on this principle, termed sonomicrometers, have been used to determine the dimensions of a variety of internal organs and structures,

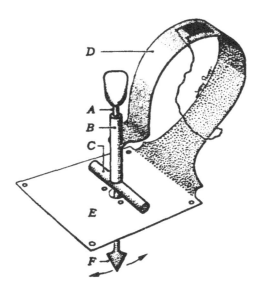

FIGURE 4. Myocardial shear transducer. A small slit cut in the myocardium with the triangular blade on the end of the shaft allows insertion of the tip. After insertion, the shaft is turned so that the corners of the blade engage the adjacent heart muscle. Axis rotation with respect to the epicardial surface displaces the thin brass shim. The resulting small bending strains are detected by strain gauges bonded to the shim. The system exhibited static linearity and frequency response essentially flat to 30 Hz. Large, rapidly changing shearing strains were recorded during isovolumetric contraction and relaxation, increasing slightly with depth below the epicardium. The shear during ejection was small only in the control state. (From Feigl, E. O. and Fry, D. L., *Circ. Res.*, 14, 1964, 536. By permission.)

including the heart, large blood vessels, liver, and spleen.[11,21,23-25] As indicated in Figure 5, the sonomicrometer is a linear system with comparatively high resolution (< 1 mm). The crystals are pulsed to deliver short bursts of ultrasound at a repetition rate on the order of 4 kHz. Thus, the frequency response is adequate. When a pair of transducers is positioned across the major internal diameter of the ventricle, a waveform is obtained such as that shown on the top trace in Figure 6A and B.[23] This measurement, however, has its own set of shortcomings. Obviously, it is a one-plane sample of a three-dimensional structure. Furthermore, and of serious concern, the waveform is not always the same when different investigators use the instrument. The hump in late diastole in Figure 6A is probably correct and indicates the effect of atrial systole on ventricular filling. Also evident in Figure 6A is a smaller bulge occurring in early isovolumetric systole, indicating the reorientation of ventricular fibers and a pre-ejection expansion of the major diameter with thinning of the wall.[23] However, the isovolumetric expansion has not been evident in data from other investigators using approximately the same technique (Figure 6B). Thus, measurements of wall thickness and dimensions with this system are only provisionally acceptable, as are those obtained by combinations of angiography and radio-opaque markers.[26,27] We are left in the position of having a general idea of what goes on, but the details of the sequence must still be considered as uncertain. The primary cause is the limitation of our measurement capability. We need better approaches with higher resolution and greater ease of placement.

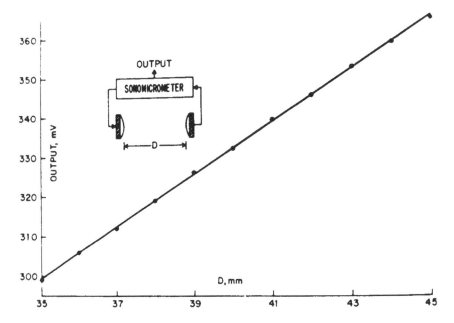

FIGURE 5. Sonomicrometer. The distance (D) between transducing elements vs. instrument output was linear. Frequency response, limited only by the pulse repetition rate, exceeds 400 Hz. The convex lenses on the surface of the piezoelectric elements diverge the ultrasonic beam to widen the sound field slightly. Any significant axial misalignment causes loss of signal. (From Kardon, M. B., Horwitz, L. D., and Bishop, V. S., in *Chronically Implanted Cardiovascular Instrumentation,* McCutcheon, E. P., Ed., Academic Press, New York, 1973, 118. By permission.)

While we are discussing ultrasound applications, there is an optical system for visualizing the ultrasonic beam that has not been noted by others in this Workshop. Questions of the detailed path of the ultrasonic beam axis arise periodically in the course of constructing and testing ultrasonic transducers. Figure 7 shows a system we used at the University of Kentucky employing a laser to visualize the ultrasonic radiation.[28] It is relatively easy and inexpensive to set up, and we found it to be very valuable in answering some of the questions about the location of the ultrasonic beam. The transducer is placed in the fish tank in the center of the optical axis of the system. Images can be recorded on Polaroid® film and viewed quickly. The effect on the beam distribution when an obstruction is placed in the path can be seen very clearly. The resolution of this system is quite high, so it is also very easy to verify directly the detailed properties within the beam.

## C. Force and Dimension: *In Situ* Veins and Arteries

Moving on to consideration of *in situ* arterial and venous characteristics, it is well known that there are very marked variations in the size and shape of the veins as seen in qualitative data. The sonomicrometer or ultrasonic transit time device can give a general indication of venous dimension. For example, changes in superior vena cava diameter can be depicted very nicely (Figure 8).[29] Others have also attempted to track venous dimensions with devices placed inside the vessels,[30] but the approaches available really do leave a great deal to be desired, and better ones in multiple dimensions are needed.

In the arterial system, devices have been placed internally, such as an inductively coupled transducer (Figure 9).[31,32] The deflecting loops placed in the vessel move with

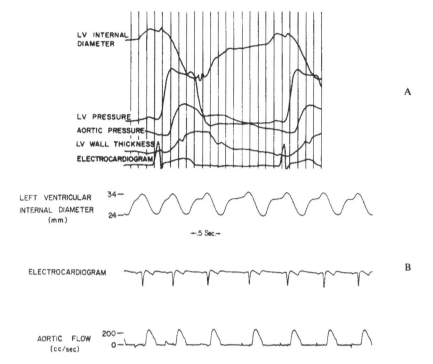

FIGURE 6. (A) High speed recording illustrating the temporal relationships of left ventricular (LV) internal diameter, LV and aortic pressures, LV anterior wall thickness, and the electrocardiogram of a chronically instrumented, conscious dog. LV internal diameter was obtained with ultrasonic transducers placed on the anterior and posterior endocardial surfaces just below the mitral annulus. LV wall thickness was detected with an adjacent epicardial transducer. The isovolumetric segment of systole is shaded. In 18 of 22 dogs, the internal diameter bulge coincident with atrial contraction was followed by a second, smaller bulge in early isovolumetric systole. LV wall thickness decreased at each of the two occurrences of volume expansion. The changes in wall thickness with placement by transmural puncture were one half of those obtained when the transducers were sewn to the endocardium under direct visualization, with the wires brought out through the ventricular apex rather than through the ventricular wall. (From Guntheroth, W. G., *J. Appl. Physiol.*, 36, 308, 1974, 310. By permission.) (B) Record of left ventricular internal diameter, electrocardiogram, and aortic flow from a chronically instrumented, resting, conscious dog. The sonocardiometer and the transmural puncture implantation techniques were similar to those used in (A), but the isovolumetric bulge is not evident. (From Horwitz, L. D. and Bishop, V. S., in *Chronically Implanted Cardiovascular Instrumentation*, McCutcheon, E. P., Ed., Academic Press, New York, 1973, 310. By permission.

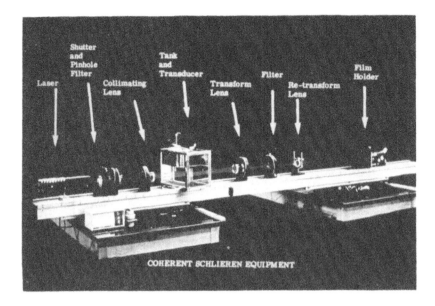

FIGURE 7. A simplified coherent Schlieren system for examination of ultrasonic fields. The 1 mW laser beam is collimated by two lenses and shone through the tank of water in which the transducer is placed. Due to the finite tank geometry, the ultrasonic radiation produces a standing wave pattern which is essentially a periodic refractive index variation within the tank. The refractive index variation acts like a diffraction grating, producing a regular diffraction pattern appearing at one focal length from the third adjacent lens. After filtering, the fourth lens provides the inverse transform for direct visualization and photographic recording of the standing waves corresponding directly to the spatial distribution and frequencies of the ultrasonic energy.

the changes in wall dimension. Data of this type indicate a 4 to 5% expansion of the normal artery with each pulse with a waveform very close to that of pressure. Similar results were reported with a noninvasive ultrasonic echo tracking system.[33] However, it is not all that difficult to get arterial deformations easily exceeding the smaller changes occurring with the normal pulse. For instance, during the process of inflating a cuff placed around the arm, the lumen of the brachial artery will vary through the complete range of fully closed to completely open (Figure 10).[34] These data are from the limbs of dogs and were obtained by casting techniques, verified and supported by radiographic findings. One would suspect that an array of the proper dimension and force transducers, relating the applied pressure to the vascular deformation, could make it possible to assess mechanical properties of any large limb artery to determine its normality. The results shown in Figure 10, although obtained from dogs, are indicative of the possibilities if there were better approaches to getting the data.

## D. Pressure

Extending our considerations of force and its distribution to that of the hydraulic force per unit area, or pressure, Greber and Ko have given indications of bright prospects on the horizon. However, the mainstay of pressure sensing is still the flat diaphragm coupled to the ubiquitous strain gauge, as shown in Figure 11A.[35] Despite the limitations of fluid-filled catheter systems, externally positioned transducers continue to be the most commonly utilized technique for both central and peripheral circulatory measurements, especially in these days when cost is of paramount consideration. Very small, directly coupled catheters and glass pipettes have been introduced into the microcirculatory vessels for mean and phasic pressure determinations, but high surface

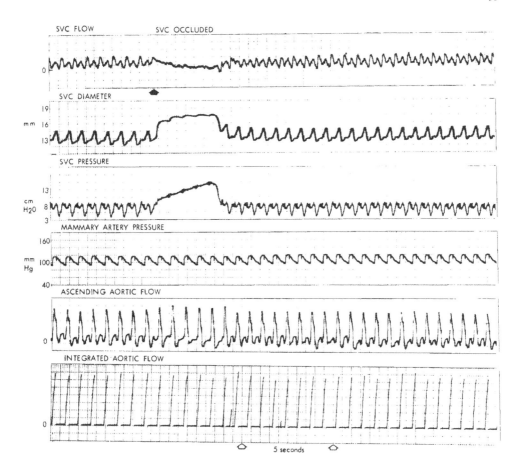

FIGURE 8.   Pulsed ultrasonic venous flow and dimension measurements in a chronically instrumented dog. Phasic flows and the effect of brief occlusion are shown for superior vena cava flow (SVC Flow), caval diameter (SVC diameter), pressure in the SVC and mammary artery, and aortic flow and stroke volume. Note the postocclusion period of constantly rising SVC pressure without further diameter increases. (From Guntheroth, W. G., in *Chronically Implanted Cardiovascular Instrumentation*, Mc-Cutcheon, E. P., Ed., Academic Press, New York, 1973, 73. By permission.)

tension limits their use for phasic pressures at arteriolar and capillary dimensions.[36,37] Phasic measurements in arterioles, capillaries, and venules have been obtained with a different principle by detecting the impedance of an electrolyte at the tip of a micropipette and measuring the pressure fed back to displace the fluid as needed to keep the impedance constant.[38,39] Of course, these are quite fragile and strictly short-term devices. While I am discussing the microcirculation, access by direct visualization is promising.[40,41] Structure and function can be examined by transillumination of the nail bed and by reflected light from the retina and skin.

When we can position the sensing element within the vasculature, we have a new dimension for obtaining high waveform fidelity applicable for either acute (manometer-tipped catheter) or very long-term (several years) data extraction as I reported at the previous meeting on pressure transducers.[42] Dogs at NASA Ames Research Center have tolerated these devices for periods exceeding 7 years. Although data were not obtained successfully for that period of time, very long-term monitoring is certainly feasible. Figure 11B shows a variety of the existing single-element, diaphragm-coupled "semiconductor" devices. Although they are of great value, they still remain very expensive, may have high drift rates, and long-term calibrations are difficult. We are

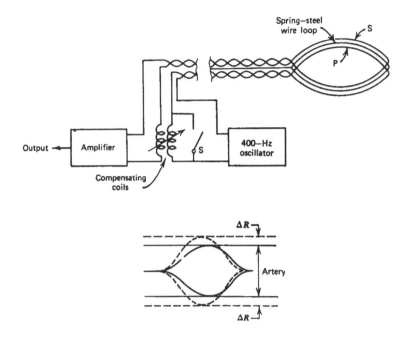

FIGURE 9.    Intravascular dimension measurement gauge based on the mutual inductance principle. The loop on the end of a catheter is introduced peripherally and routed to the measurement site where it expands to follow the internal dynamic dimension fluctuations. The 400 Hz signal in loop P induces in S a voltage dependent on loop area. The induced voltage depends on the log of the arterial diameter with a resolution of ±0.3 mm. (From Cobbold, R. S. C., *Transducers for Biomedical Measurements: Principles and Applications,* John Wiley & Sons, New York, 1974, 142. By permission.)

looking forward to solutions to these problems, yet in spite of all this, the choice of suitable devices for pressure measurement is the widest of any of the variables.

## III. VOLUME, FLOW, AND MOTION

### A. Volume: Single Dimension

A very appealing approach because of its simplicity is to estimate volume, and volume changes, from a single dimension. This has been attempted with some success. A mercury-in-rubber strain gauge positioned in an epicardial arc length at the ventricular equator was used to determine left ventricular volume and the volume change with ejection.[20] High correlations were found with stroke volume estimates obtained from the area under the aortic flow waveform. By the way, their epicardial waveform also did not show the extra pre-ejection diameter bulge discussed earlier.[23] Electromagnetically coupled coils have been used frequently, but if such coils are not fixed axially, accuracy is limited by the very great difficulty of maintaining alignment of the coils throughout the cycle.[16,20] In the case of nonconstrained electromagnetically coupled coils, errors can be on the order of 20 to 30%.[20] Thermodilution is of reported value for ventricular volumes, but I have had a lot of trouble getting satisfactory reproducibility in the comparatively small heart of dogs, as have others.[43] On the other hand, successful applications have been reported in animals and humans.[44-46] Moreover, a combined thermodilution and hydrogen concentration detection system has been used to estimate left ventricular mass from samples at the coronary sinus.[47]

97

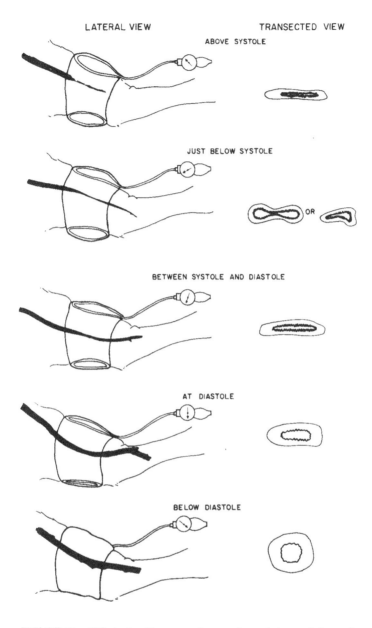

LATERAL VIEW          TRANSECTED VIEW

ABOVE SYSTOLE

JUST BELOW SYSTOLE

BETWEEN SYSTOLE AND DIASTOLE

AT DIASTOLE

BELOW DIASTOLE

FIGURE 10. Effect of cuff compression on size and shape of the canine brachial artery. The intraarterial casts were made with a polyester resin which solidified in about 10 min, so each example is a separate experiment. The appearance on cineangiography was similar, with the intraarterial contrast material exhibiting a pulsating tapered shape and reduced internal dimensions. Unloading of the vessel by the decreased transmural pressure during cuff inflation leads to varying degrees of constriction and collapse. The characteristics of these changes would be expected to vary with the vessel compliance. (From Abel, F. L. and McCutcheon, E. P., *Cardiovascular Function: Principles and Application*, Little Brown, Boston, 1979, in press. With permission.)

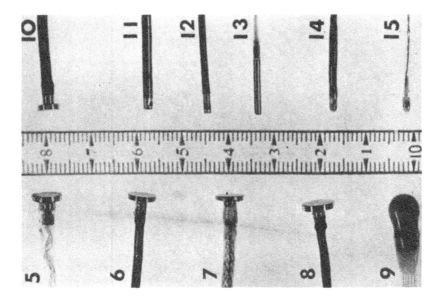

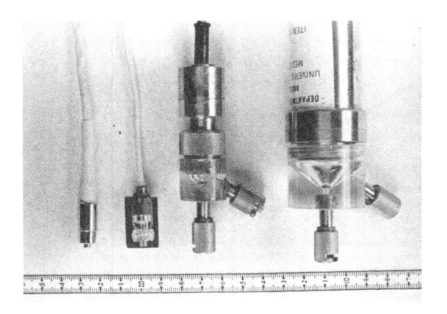

FIGURE 11.    Many types of pressure transducers are available. (A) Manometers for connection to catheters. (B) Manometers for intravascular placement.

In addition to the ultrasonic transmission data cited earlier, direct and transcutaneous application of the reflectance mode, or echo-location, for calculation of volumes is now widespread in animals and humans. The techniques and implications of ultrasonic imaging are emphasized in the paper by Donald Baker. As has been shown, even a single echo signal directed across a major diameter can produce useful data even in fairly difficult circumstances.[48-52] For instance, in humans exposed to lower body negative pressure (LBNP), the diastolic dimension is observed becoming much smaller and then recovering with release of the LBNP (Figure 12).[52] The data obtained from echocardiography have relatively high density, and it takes a good deal of expertise to develop consistency in interpreting such information. There are also technical limitations having to do, for instance, with difficulties of determining anterior wall thickness and the difficulty of tracking both dimensions and wall thickness in the same plane at the same time. Perhaps, progress with the development of the ultrasonic arrays will provide the order of magnitude improvement that I think is required to make ultrasonic dimension detection viable for truly quantitative information. As part of this validation process, and to permit recording from unrestrained animals, work on a totally implantable system is underway at NASA Ames Research Center.[53] A transducer for this system is shown in Figure 13. It is essentially a truncated version of a standard clinical unit encased in silicone rubber for attachment flexibility. This transducer is powered by inductive coupling. Therefore, the basic system can be implanted in the animal with power supplied through an external transmitting coil placed on the skin over an internal pick-up coil. The pulses generated in the implanted system are transmitted out through a second coil, received externally, boosted a bit, and supplied to the input of a standard echocardiograph. The 3 MHz information is transmitted right through the skin of the animal to the receiving coil. The device will be useful for analyzing the properties and problems of ultrasonic measurement techniques, and such animal preparations can provide validation by comparison with other approaches.

## B. Volume: Area

Imaging techniques continue to be very appealing. A significant measurement applicable to the intact, closed-chest circulation of experimental animals and a mainstay in human circulatory studies is radiologic contrast angiography.[26,27,54,55] Radiopaque material injected into the circulation permits visualization of internal volumes in single or multiple planes. By use of movie cameras, sample rates can be 60 Hz or better for intrabeat data. Drawbacks are (1) the great expense of equipment acquisition and maintenance, (2) the radiation hazard for subjects and technical personnel, (3) consistent overestimation of the internal dimension and wall thickness, (4) and the requirement for calibration by comparison to volumes obtained with another technique, such as casts of the post-mortem cardiac chambers.[27,55] It is valuable for demonstrating the reason for some of the persistent shortcomings of one-dimensional sampling of ventricular function. Despite the difficulties, for a considerable period of time angiography will undoubtedly provide a standard method against which other techniques can be compared and quantified.

The functional understanding added by angiography can be illustrated by an example from examination of the left ventricular chamber. Determination of the complex systolic cardiac contraction sequence and pattern (Figure 14) has analogies with the sensing of pressure as discussed earlier by Dr. Greber. From the outline of a dog ventricle beginning at end-diastole and moving into end-systole, it is apparent that the contraction is not very symmetrical or synchronous. The sequence requires 60 to 80 msec to develop, and as apparent even in the normal heart, it can not be a truly isotropic, homogeneous sequence. When the volume available to the heart is reduced by

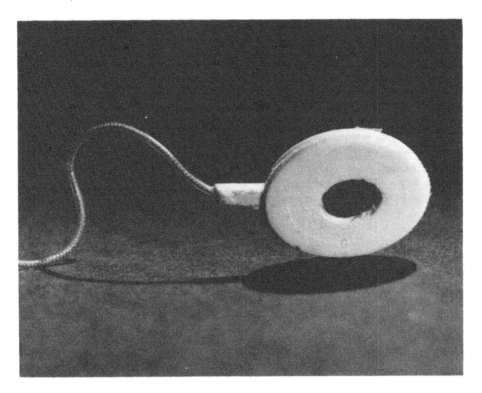

FIGURE 13.   Transducer for use with an implantable echocardiometer, a technique under development for obtaining ultrasonic cardiac dimension information from active animals.

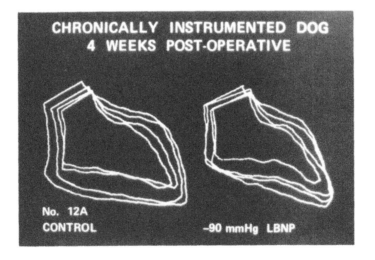

FIGURE 12.   Representative left ventricular ultrasonic echogram from a normal human subject for control, during 60 mmHg lower body negative pressure (LBNP), and post LBNP recovery periods. Sample axis was determined by positioning the beam just below the mitral valve plane with the transducer in the fourth intercostal space at the left sternal border of the supine subject. Systolic internal dimension = S, and diastolic internal dimension = D. (Courtesy of Popp, R. L., Stanford University Medical Center. By permission.)

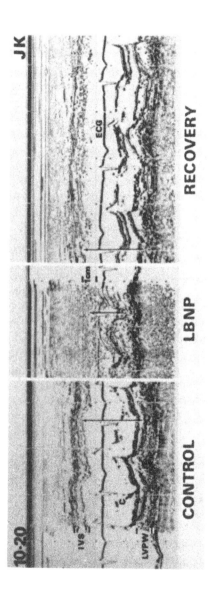

FIGURE 14. Progressive changes from end-diastolic to end-systolic configuration of the internal left ventricular outline determined by cineangiocardiography in the dog. (A) Control. (B) After reduction of effective circulating blood volume with 90 mmHg, lower body negative pressure (LBNP). Decreased movement of the aortic valve plane after LBNP was a characteristic change with chamber volume reduction.

applying a circulatory stress to the animal such as lower body negative pressure, the resulting small ventricular size is associated with additional changes in the contraction pattern. Thus, sampling from a single region is severely limiting for the shape change and cardiac wall strain vary widely in different areas. Here is a good example of the need for smaller tranducers, not just to have small ones available, but to allow placement of a transducer array which can document the distributed characteristics of the contraction cycle at very high sample rates, particularly in the chronically instrumented animal.

## C. Flow

Hot-film flow velocity sensing was described elegantly by Dr. Nerem and Doppler flow by Donald Baker. When we can fix the vessel cross-sectional area within a rigid, but not constricting, encircling element or cuff, then we can obtain volume flow from a measurement device incorporated in the cuff. In the case of the macrocirculation, the electromagnetic flowmeter remains the standard and is usable in vessels down to about 1 to 2 mm in diameter.[56] Ultrasonic Doppler techniques are readily available now in a directional configuration and continue to improve and be more fully understood. Specific advantages are low power consumption and lightweight transducers. At NASA Ames Research Center, work is in progress to improve these transducers. Figure 15* shows a prototype design for use on a small vessel such as a coronary artery with the prospect of the same kind of long-term utility in animals that is found with electromagnetic systems. The continuous wave (CW) Doppler with appropriately sized sensing elements can offer an advantage over pulsed systems for measurement of volume flow. With the CW Doppler, the entire three-dimensional flow channel is sampled.[57] Pulsed systems require integration of an accurate profile obtained in a single plane of the flow channel. The present state of the art seems to me to limit confidence in the accuracy of the pulsed approach for general use to determine volume flow. Multielement transducers can improve the confidence level a bit. There are three elements in the transducer portion of the device shown in Figure 16 which Stanford University investigators implanted on the mitral valve ring of dogs to study the hemodynamics of artificial valve placement.[58] By measuring the velocity profile in multiple axes and combining the results, a quite reasonable mean velocity through the valve plane was obtained. A brief period of negative flow was observed, apparently because the transducer was located slightly above the valve plane, so presumably, there is a transient period of valve deformation at closure producing a brief surge of retrograde flow (mitral closing volume).

In both the macro- and microcirculations, dye dilution techniques are indispensable for flow measurements. These are sensed with physical means, but are outside the scope of this presentation. Noninvasive computation of flow velocity in directly visualized microcirculatory beds was referred to earlier.[40,41]

## D. Motion

For motion analysis, the sensing options pertinent to the cardiovascular system apply first to the internal organs and structures; second, they apply to the total body motion induced by cardiovascular activity; and finally, they apply to the input from dynamic environments such as sustained or oscillatory acceleration. The current availability of a number of accelerometer types and designs facilitates such studies.[59] Figure 17 shows an example of the classic configuration, a mass-loaded cantilevered beam with a strain gauge to detect beam deflection. These accelerometers have been implemented in single

---

* Figure 15 appears on the color insert that follows page 88.

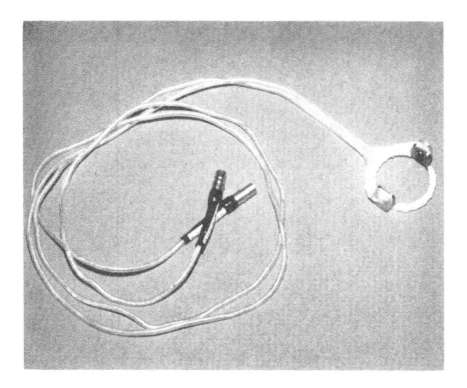

FIGURE 16.   A three-element pulsed ultrasonic Doppler transducer sewn into the mitral valve ring for study of valve flow dynamics in chronically instrumented animals.

or multiaxis configurations, and they can be incorporated into telemetry systems such as that shown in Figure 17B which is of a dimension, weight, and size allowing its use in animals as small as 10 kg or so, on the order of a macaque monkey.[60] Hopefully, these somewhat bulky, limited sensors can be replaced with some of the accelerometers that Dr. Angell described. The intrinsic ballistocardiogram is one example of the external use of accelerometers. When accelerometers are placed to determine how the body or organs are moving, the magnitude of imposed stresses such as centrifugation and vibration can be determined more precisely.

## IV. BIOPOTENTIALS

The area of biopotentials is certainly a topic unto itself. Volumes have been written on it. The concern is with both intracellular potentials and the net electrical field as exemplified by the traditional model of the electrocardiogram. One of the developments taking place in electrocardiography which is somewhat encouraging from the standpoint of acceptance of the array concept is the use of a large number of electrocardiographic electrodes with intensive mapping under computer control to improve resolution and describe the internal source characteristics much more accurately.[61]

## V. COMMENTS ON DEVICE CRITERIA

I'd like to make a few comments about the criteria for use of devices in the circulation. Table 1 lists typical cardiovascular variables, typical kinds of sensors used, and a representative frequency bandwidth for each of the variables. Note the relationship

CROSS-SECTION OF SINGLE-AXIS ACCELEROMETER

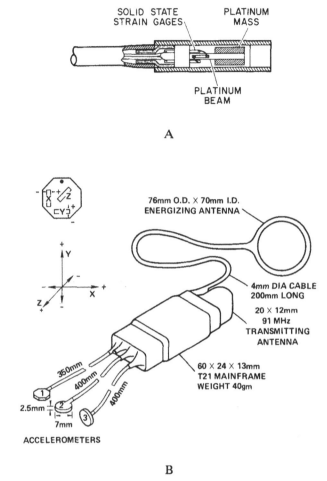

A

B

FIGURE 17. Implantable accelerometer transducers. (A) Construction principle. (B) Configuration of an inductively powered, implantable telementry system adapted for sampling of acceleration at multiple sites and axes.

between the required bandwidth and the sampling site. It is interesting that the frequency content of most cardiovascular signals is within the DC to 200 Hz region. By many people's standards, these are very low-frequency data. The minimum bandwidth requirements are one source of optimism for obtaining improved sensor systems. In fact, there are many situations where a bandwidth higher than that really necessary causes more problems than it solves. An example was a pressure transducer that was moved straight from wind tunnel to biological applications and had a frequency response of 200,000 Hz.[35] Just breathing on it set up transient resonances that shattered the diaphragm. We want to be sure that we record the variable appropriately, but we do not want extraneous information, and we do not want to introduce unnecessary fragility or noise.

TABLE 1

Representative Characteristics of Cardiovascular Measurements

| Physiologic variable | Sensor | Data bandwidth (Hz) |
|---|---|---|
| Pressure | Strain gauge on diaphragm | DC—100 |
| Flow | Electromagnetic coil or ultrasonic crystals | DC—100 |
| Distance | Ultrasonic crystals | DC—100 |
| Strain/Shear | Strain gauge on beam | DC—100 |
| Velocity | Thermistor or ultrasonic | DC—100 |
| Acceleration | Strain gauge on mass-loaded beam | DC—200 |
| Temperature | Thermistor | DC—1 |
| Biopotentials | | |
| ECG | Contact electrodes | 0.01—250 |
| Intracellular | Glass microelectrodes | DC—10,000 |
| Heart sounds | Microphone | 5—2,000 |

To extend these comments a bit, I have to get in a plug for what I have heard little of at the previous meetings and at this one, and that is the aspect of biocompatibility. Dr. Angell did mention the Parylene® coating process for which I have great hopes. However it is done, we must minimize the tissue reactivity to any implant. Particularly in the case of devices within the circulation, the thrombogenic potential can spell the difference between success and failure regardless of its performance on the bench. An associated limitation is the crudeness of our techniques for attaching the various devices. One can get by with sutures through holes, but there ought to be a better method.

## VI. CONCLUSIONS

In closing, let me summarize some of my concerns about the considerations of getting higher quality cardiovascular data. First of all, I think it is clear that the system complexities and subtleties require assessment of its distributed characteristics in both time and space. Biological structures, as a rule, are not homogeneous, and the geometry is seldom plane. The broad category of the solution to such problems is the use of arrays of many types and configurations. I find this prospect very appealing and challenging and one we appear to be on the threshold of achieving. We have seen its beginning in integrated and multiparameter transducing systems such as Baker's duplex scanner and other similar approaches. This is the sort of thing that miniaturization allows, conveying an advantage beyond just having a small device. We are, above all, after quantitative data, not just qualitative. We have to calibrate our instrumentation, and we want to put accurate numbers on as many of these critical variables as we can. I expect that (as has been true in the past) as these devices evolve and as validation is conducted, the animal will always be involved in the loop. The whole point of doing all the animal work is to develop something applicable to the human situation. Hopefully, we are in a transitional situation as far as the current state of applications is concerned, and I anticipate that it probably will not be long before the overview I presented is regarded as quaint and primarily of historical value.

# REFERENCES

1. **Jacobs, H. K., McConnell, D. P., Rowley, B. A., and South, F. E.**, A force transducer for cardiac muscle strips, *J. Appl. Physiol.*, 35, 436, 1973.
2. **Hamrell, B. B., Panaanan, R., Trono, J., and Alpert, N. R.**, A stable, sensitive, low-compliance capacitance force transducer, *J. Appl. Physiol.*, 38, 190, 1975.
3. **Murgo, J. P., Cox, R. H., and Peterson, L. H.**, Cantilever transducer for continuous measurement of arterial diameter in vivo, *J. Appl. Physiol.*, 31, 948, 1971.
4. **Cox, R. H.**, Three-dimensional mechanics of arterial segments in vitro: methods, *J. Appl. Physiol.*, 36, 381, 1974.
5. **Cox, R. H.**, Determination of series elasticity in arterial smooth muscle, *Am. J. Physiol.*, 233, H248, 1977.
6. **Huntsman, L. L., Day, S. R., and Stewart, D. K.**, Nonuniform contraction in the isolated cat papillary muscle, *Am. J. Physiol.*, 233, H613, 1977.
7. **Pinto, J. G. and Win, R.**, Nonuniform strain distribution in papillary muscles, *Am. J. Physiol.*, 233, H410, 1977.
8. **Cox, R. H.**, Effects of age on the mechanical properties of rat carotid artery, *Am. J. Physiol.*, 233, H256, 1977.
9. **Learoyd, B. M. and Taylor, M. G.**, Alterations with age in the viscoelastic properties of human arterial walls, *Circ. Res.*, 28, 278, 1966.
10. **Cotten, M. de V. and Bay, E.**, Direct measurement of changes in cardiac contractile force: relationship of such measurements to stroke work, isometric pressure gradient and other parameters of cardiac functions, *Am. J. Physiol.*, 187, 122, 1956.
11. **Cobbold, R. S. C.**, *Transducers for Biomedical Measurements: Principles and Applications*, John Wiley & Sons, New York, 1974.
12. **Feigl, E. O., Simon, G. A., and Fry, D. L.**, Auxotonic and isometric cardiac force transducers, *J. Appl. Physiol.*, 23, 597, 1967.
13. **Hefner, L. L., Sheffield, L. T., Cobbs, G. C., and Klip, W.**, Relation between mural force and pressure in the left ventricle of the dog, *Circ. Res.*, 11, 654, 1962.
14. **Feigl, E. O. and Fry, D. L.**, Myocardial mural thickness during the cardiac cycle, *Circ. Res.*, 14, 541, 1964.
15. **Patel, D. J., Janicki, J. S., and Carew, T. E.**, Static anisotropic elastic properties of the aorta in living dogs, *Circ. Res.*, 25, 765, 1969.
16. **McHale, P. A. and Greenfield, J. C., Jr.**, Evaluation of several geometric models for estimation of left ventricular circumferential wall stress, *Circ. Res.*, 33, 303, 1973.
17. **Lewartowski, B., Sedek, G., and Okolska, A.**, Direct measurement of tension within left ventricular wall of the dog heart, *Cardiovasc. Res.*, 6, 28, 1972.
18. **van der Meer, J. J., Jageneau, A. H. M., Elzinga, G., Grondelle, R. V., and Reneman, R. S.**, Changes in myocardial wall thickness (MWT) in open chest dogs, *Pfluegers Arch.*, 340, 35, 1973.
19. **Feigl, E. O. and Fry, D. L.**, Intramural myocardial shear during the cardiac cycle, *Circ. Res.*, 14, 536, 1964.
20. **Wilson, G. J. and Bergel, D. H.**, Continuous measurement of left ventricular volume using a single dimensional transducer: a comparison of two techniques in open chested dogs, *Cardiovasc. Res.*, 9, 327, 1975.
21. **Kardon, M. B., Horwitz, L. D., and Bishop, V. S.**, The ultrasonic pulse transit time technique, in *Chronically Implanted Cardiovascular Instrumentation*, McCutcheon, E. P., Ed., Academic Press, New York, 1973, 117.
22. **Rubinstein, M. L., Carlson, C. J., and Rapaport, E.**, Improved signal processor for the ultrasonic dimension gauge, *Am. J. Physiol.*, 233, H322, 1977.
23. **Guntheroth, W. G.**, Changes in left ventricular wall thickness during the cardiac cycle, *J. Appl. Physiol.*, 36, 308, 1974.
24. **Mahler, F., Covell, J. W., and Ross, J., Jr.**, Systolic pressure-diameter relations in the normal conscious dog, *Cardiovasc. Res.*, 9, 447, 1975.
25. **Renkin, J. S., McHale, P. A., Arentzen, C. E., Ling, D., Greenfield, J. C., Jr., and Anderson, R. W.**, The three-dimensional dynamic geometry of the left ventricle in the conscious dog, *Circ. Res.*, 39, 304, 1976.
26. **Leshin, S. J., Wildenthal, K., Mullins, C. B., and Mitchell, J. H.**, Left ventricular dimensions determined by biplane cinefluorography of chronically implanted radiopaque markers: critique of the method, in *Chronically Implanted Cardiovascular Instrumentation*, McCutcheon, E. P., Ed., Academic Press, New York, 1973, 123.
27. **Mirsky, I., Ghista, D. N., and Sandler, H.**, *Cardiac Mechanics: Physiological, Clinical, and Mathematical Considerations*, John Wiley & Sons, New York, 1974.

28. Adams, R. N., McCutcheon, E. P., and Stanifer, R. R., Coherent Schlieren and ultrasonics, Proc. 8th Int. Conf. Eng. Med. Biol., July, 1969, Session 10, p. 2.

29. Guntheroth, W. G. and Chakmakjian, S., Active changes in tone in the canine vena cava, Circ. Res., 28, 554, 1971.

30. Reddy, R. R. V., Iwazumi, T., and Moreno, A. H., A catheter-tip probe for dynamic cross-sectional area measurement, in Chronically Implanted Cardiovascular Instrumentation, McCutcheon, E. P., Ed., Academic Press, New York, 1973, 149.

31. Baan, J., Iwazumi, T., Szidon, J. P., and Noordergraaf, A., Intravascular area transducer measuring dynamic local distensibility of the aorta, J. Appl. Physiol., 31, 499, 1971.

32. Kolin, A. and Culp, G. W., An intra-arterial induction gauge, IEEE Trans. Biomed. Eng., BME-18, 110, 1971.

33. Hokansen, D. E., Mozersky, D. J., Sumner, D. S., and Strandness, D. E., Jr., A phase-locked echo tracking system for recording arterial diameter changes in vivo, J. Appl. Physiol., 32, 728, 1972.

34. McCutcheon, E. P. and Rushmer, R. F., Korotkoff sounds: an experimental critique, Circ. Res., 20, 149, 1967.

35. McCutcheon, E. P., Evans, J. M., and Stanifer, R. R., Direct blood pressure measurement: gadgets vs. progress, Anesth. Analg., 51, 746, 1972.

36. Raper, A. J. and Levasseur, J. E., Accurate sustained measurement of intraluminal pressure from the microvasculature, Cardiovasc. Res., 5, 589, 1971.

37. Intaglietta, M., Pressure measurements in the microcirculation with active and passive transducers, Microvasc. Res., 5, 317, 1973.

38. Fox, J. R. and Wiederhielm, C. A., Characteristics of the servo-controlled micropipet pressure system, Microvasc. Res., 5, 324, 1973.

39. Wiederhielm, C. A., Woodbury, J. W., Kirk, S., and Rushmer, R. F., Pulsatile pressures in the microcirculation of frog's mesentery, Am. J. Physiol., 207, 173, 1964.

40. Butti, P., Intaglietta, M., Reiman, H., Hollinger, C. H., Bollinger, A., and Anliker, M., Capillary red blood cell velocity measurements in human nail fold by videodensitometric method, Microvasc. Res., 10, 1, 1975.

41. Fagrell, B., Fronek, A., and Intaglietta, M., A microscope-television system for studying flow velocity in human skin capillaries, Am. J. Physiol., 233, H318, 1977.

42. McCutcheon, E. P., Survey of present applications for indwelling pressure transducer systems and experiences with cardiovascular implants, in Indwelling and Implantable Pressure Transducers, Fleming, D. G., Ko, W. H., and Neuman, M. R., Eds., CRC Press, Cleveland, 1977, 47.

43. Magrini, F. and Tarazi, R. C., New approach to local thermodilution: use of pigtail catheters to avoid basic difficulties, Cardiovasc. Res., 11, 576, 1977.

44. Ganz, W., Donoso, K., Marcus, H. S., Forrester, J. S., and Swan, H. J. C., A new technique for measurement of cardiac output by thermodilution in man, Am. J. Cardiol., 27, 392, 1971.

45. Fronek, A., Thermodilution in chronic experimentation, in Chronically Implanted Cardiovascular Instrumentation, McCutcheon, E. P., Ed., Academic Press, New York, 1973, 77.

46. Warren, D. J., A thermistor probe for measurement of blood temperature and cardiac output in small animals, Cardiovasc. Res., 8, 566, 1974.

47. Crawford, D. W. and Barndt, R., Jr., In-vivo estimation of left ventricular mass: a double indicator technique using a single sensing catheter, Cardiovasc. Res., 8, 707, 1974.

48. Popp, R. L. and Harrison, D. C., Ultrasonic cardiac echocardiography for determining stroke volume and valvular regurgitation, Circulation, 41, 493, 1970.

49. Troy, B. L., Pombo, J., and Rackley, C. E., Measurement of left ventricular wall thickness and mass by echocardiography, Circulation, 45, 602, 1972.

50. Belenkie, I., Nutter, D. O., Clark, D. W., McCraw, D. B., and Raizner, A. E., Assessment of left ventricular dimensions and function by echocardiography, Am. J. Cardiol., 31, 755, 1973.

51. Kisslo, J. A., von Ramm, O. T., and Thurstone, F. L., Dynamic cardiac imaging using a focused, phased-array ultrasound system, Am. J. Med., 63, 61, 1977.

52. Popp, R. L., Brown, O. R., Filly, K., and Sandler, H., Serial echocardiographic measurement of left ventricular dimensions during subacute and acute circulatory stress, Proc. 19th Annu. Conf. Am. Inst. Ultrasound in Med., Seattle, October 5 to 10, 1974, A-23.

53. Kojima, G. K., McCutcheon, E. P., Carlson, E. L., and Lee, R. D., A miniature implantable echo ultrasonometer, presented at Proc. 3rd Int. Symp. Biotelemetry, Pacific Grove, Calif., May 17 to 20, 1976.

54. Harrison, D. C., Sandler, H., and Miller, H. A., Eds., Cardiovascular Imaging and Image Processing, The Society of Photooptical Instrumentation Engineers, Palos Verdes Estates, Calif., 1975, 72.

55. Laks, M. M., Garner, D., Beazell, J., and Piscatelli, J., A new method for internal calibration of left ventricular cineangiography, Am. J. Physiol., 232, H434, 1977.

56. **White, S. W., McRitchie, R. J., and Porges, W. L.**, A comparison between thermodilution, electromagnetic and Doppler methods for cardiac output measurement in the rabbit, *Clin. Exp. Pharm. Physiol.*, 1, 175, 1974.

57. **McCutcheon, E. P. and Freund, W. R.**, Theoretical and experimental analysis of the accuracy of CW Doppler flow measurement with implantable transducers, presented at Proc. 29th Annu. Conf. Eng. Med. Biol., Boston, Mass., November 6 to 10, 1976.

58. **Freund, W. R., Salomon, N. W., and Torman, H. A.**, A technique for measurement of mitral flow hemodynamics in chronically instrumented animals, presented at Proc. 21st Annu. Conf. Am. Inst. Ultrasound in Med., San Francisco, August 3 to 7, 1976, 107.

59. **Sandler H., Fryer, T. B., Westbrook, R. M., and Konigsberg, E.**, Miniature implantable accelerometers, in *Chronically Implanted Cardiovascular Instrumentation*, McCutcheon, E. P., Ed., Academic Press, New York, 1973, 165.

60. **McCutcheon, E. P., Miranda, R., Fryer, T. B., and Carlson, E. L.**, An inductively powered implantable multichannel telemetry system for cardiovascular data, presented at Proc. 3rd Int. Symp. Biotelemetry, Pacific Grove, Calif., May 17 to 20, 1976.

61. **Barr, R. C., Ramsey, M., and Spach, M. S.**, Relating epicardial to body surface potential distributions by means of transfer coefficients based on geometry measurements, *IEEE Trans. Biomed. Eng.*, BME-24, 1, 1977.

Chapter 8

# INTEGRATED CIRCUIT FORCE TRANSDUCERS FOR CHRONIC IMPLANTATION

Thomas S. Nelsen,* Timothy A. Nunn, Holde H. Muller, and James B. Angell

## TABLE OF CONTENTS

* Supported by National Institute of Child Health and Human Development Contract # 70-2257.

# I. INTRODUCTION

As an integral part of a long-term program to develop microminiature implantable transducers for chronic physiologic studies in intact animals and man, an integrated circuit force transducer has been developed. Corollary devices for measuring electrical activity and pressure have also been made, but will not be discussed in this paper. The design goals for the force transducer were as follows:

1.  A 1-mm basic transducer span enabling study of the smooth muscle of the rabbit oviduct and other thin, small, smooth muscle structures without interference with tissue function
2.  A size, shape, and composition suitable for transducer withdrawal by means of the lead wires permitting human implant during laparotomy
3.  Adequate sensitivity and long-term stability, reproducible characteristics
4.  No reaction in the tissue environment
5.  Reasonable cost
6.  Easy implantation, no cross-interference with multiple implants

The fabrication, engineering characteristics, and testing of the final design of this device, which is suitable for chronic implantation, singly or in arrays, is described in detail in the paper by J. Angell, elsewhere in this publication. This design represents the improvements suggested by repeated trials of various forms of the force transducers constructed during the past several years. It incorporates several useful features:

1.  The doped silicon force element with integral suture loops for attachment to tissue with nylon suture
2.  A lead assembly sufficiently limp so that no interference with gauge function is engendered by motion of abdominal viscera with respiration, skeletal motion, etc.
3.  Parylene® insulation ensuring nonreactivity in tissue
4.  Average gauge life of 60 days in vivo
5.  An infection-free inexpensive means for bringing the lead wires through the skin.

A detailed, illustrated brochure on implantation technique is available from the authors.

Although care must be used in handling the force transducers with surgical instruments, once in place on the tissue they are virtually indestructible (by tissue forces). Figure 1* shows the transducer element, intermediate Pt-Ir wires, and the block of silicone rubber used to insulate the wire solder joints and stabilize the lead assembly. Figure 2 shows three transducers sutured to a rabbit oviduct. The intermediate blocks are sutured to the mesotubarium.

The essential application of the force transducers has been on experiments concerned with the motility of the rabbit oviduct in various physiologic states. The final series of these studies has required over 100 transducers, and an average functional life of 60 days has been achieved. Samples of recordings taken from various experiments on the Fallopian tube of the rabbit are shown in Figure 3.

During the course of testing the force transducers, they have been attached to a variety of rabbit tissues capable of active contraction and yielded excellent recordings.

---

* Figure 1 appears on the color insert that follows page 88.

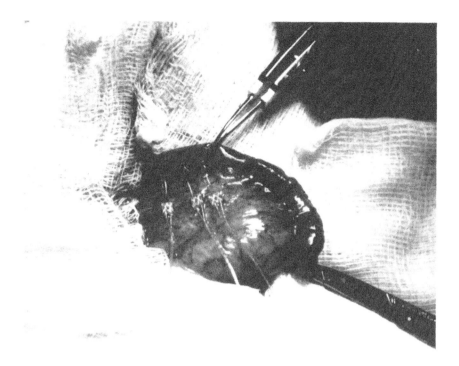

FIGURE 2. Shows three force transducers sutured to the long axis of the rabbit oviduct. The silicone rubber pads are sutured to the mesovarium.

Sample tracings from pylorus, duodenum, aorta, and uterus are shown in Figure 4. They are also suitable for use on bone, cartilage, ligaments, heart, etc., employing suitable modifications in the mode(s) of attachment.

In summary, the current design of the integrated circuit force transducer has met or exceeded the original design performance goals in extensive in vivo testing in one system, the rabbit oviduct. The devices should prove to be extremely useful in a wide variety of organ systems and experimental situations.

## II. SILICON MICROMINIATURE FORCE TRANSDUCER IMPLANTATION: SURGICAL PROCEDURE

Female mature rabbits (New Zealand White) of approximately 3 to 4 kg body weight have been used. Following an i.p. injection of 26.4 mg/kg sodium pentobarbital (Veterinary Diabutal, sterile) a mid-ventral incision is made and the reproductive tract gently exteriorized and examined.

The electrical resistance of each force transducer is checked before and during implantation. The thinner, diffused resistor portion (30μm) is more vulnerable to breakage than the rest of the force transducer (65μm). Consequently, extreme care has to be taken in handling the devices. The double-ended microminiature force transducer with its integral silicon suture loops is sewn onto the isthmus portion of the Fallopian tube with 9-0 monofilament nylon suture. Usually, the long axis of the gauge is aligned with the longitudinal muscle of the oviduct, but transverse orientation is feasible. The perforated silicone rubber block with platinum wire-copper wire solder joints encased is sutured on the mesasalpinx to hold the assembly in place and stabilize it against twisting. An additional stitch, in the form of a loose loop surrounding the lead wires, is taken in nearby fatty tissue to stabilize the lead wires and rubber block and avoid

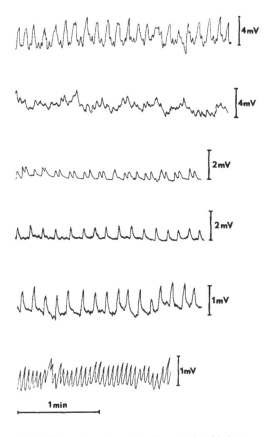

FIGURE 3.   Examples of tracings made with force transducers sutured to the rabbit oviduct in varying physiologic conditions.

subjecting them to excessive motion from the mesosalpinx (Figure 2). Great care should be taken to limit adhesions by minimizing dissection, tissue manipulation, and drying.

After gauge attachment, the leadwires are gathered into a loose bundle which is loosely looped in the abdomen to provide slack. An amputated 3 ml plastic syringe is drawn through the abdominal wall via a tapered stainless steel cannula puller (Figure 4). This requires no incision and ensures that the muscles and skin will seal tightly around the syringe barrel. Figures 5 to 9 display the rest of the procedure.

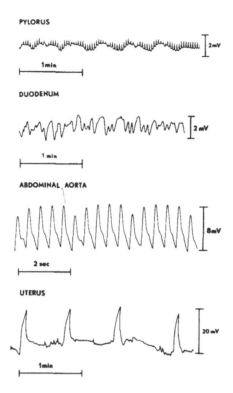

FIGURE 4. Representive recordings made
from force transducers sutured to other muscular
structures in the rabbit.

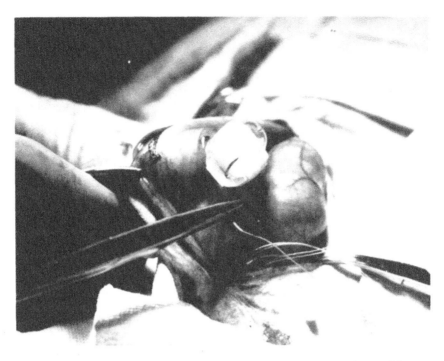

FIGURE 5.   The lumen of the plastic cannula has been previously sealed by molding into place a silicone rubber plug containing a small concentric hole and suture ears, which are subsequently sewn with a 3-0 chromic suture to the inside of the abdominal wall.

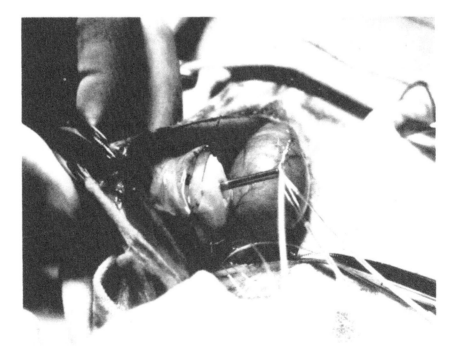

FIGURE 6.   The wires are passed out of the abdomen through the hole in the rubber plug.

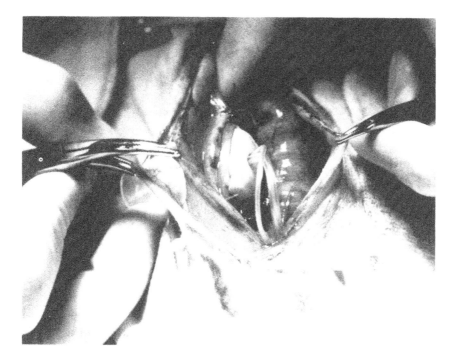

FIGURE 7. This hole is immediately sealed with catalyzed medical-grade silicone rubber which bonds to the molded rubber plug and cures *in situ.*

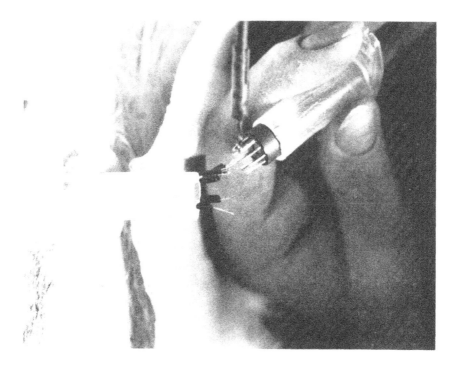

FIGURE 8. After the abdominal wound is closed in two layers, the multistranded copper wires are thermally stripped, dipped in melted beeswax, and subsequently soldered to a female multipin connector fitting snugly inside the cannula.

A                                    B

FIGURE 9.  (A and B) For mechanical protection from the rabbit's teeth, a machined collar of metal is locked onto the cannula with a screw.

## REFERENCES

1. **Nelsen, Thomas S., Nunn, Timothy A., and Angell, James B.,** Microminiature transducers for oviductal motor function, in Ovum Transport and Fertility Regulation, Harper, M. J. K., Ed., World Health Organization, Copenhagen, 1976, 75.

## DISCUSSION

**Dr. F. T. Hambrecht:** You mentioned that your longest in vivo survival was about 90 days. Could you elaborate on the failure modes that you're seeing in vivo?

**Dr. Nelsen:** The majority of the recent failures have been leakage through the insulation of the stranded wire(s) with subsequent wire failure. We formerly had trouble with bonding of the platinum wires to the pads on the force transducer. This problem has been corrected.

**Dr. Hambrecht:** Have you seen any moisture beneath the Parylene®?

**Dr. Nelson:** Only in the gauge assemblies that we believe to be improperly coated. Those with adequate thickness of Parylene®, 10 μm or more, have been satisfactory. The Parylene® coating is brought us as a stocking on the lead wires, but the Parylene® layer doesn't extend to the connector.

**Peter Jacobson:** Would you comment on the biocompatibility of siver epoxy?

**Dr. Nelsen:** I don't know that any serious investigation has been done. It is not an FDA approved material for implantation.

**Dr. S. K. Wolfson:** You mentioned the desirability of being able to extract the force transducer via the leads, and you didn't mention that in connection with this device. Does this have any such possibilities for human implantation?

**Dr. Nelsen:** We think so. We have pulled some loose, and it's quite easy to extract them with the lead wire assembly if you use the proper suture. Even finer, more fragile, sutures could be used. We have done no human implantation nor have we applied for permission to do them.

**Dr. Wolfson:** Wouldn't that lucite block, or whatever it was, with the suture for the base of the external lead be a problem in trying to extract it?

**Dr. Nelsen:** I don't think so. It is made of fully approved, implantable material (silicone rubber).

**Dr. Wolfson:** No, I mean from the point of view of tearing that loose and bringing it out.

**Dr. Nelsen:** No, the suture on the block will tear out of the tissue. A single loop suture through the hole is placed in the tissue underneath and the suture simply pulls out with traction.

**Dr. Neuman:** With regard to the propagating waves that you see along the isthmus, as you know, there are some people who say there isn't much propagation and others who say there's just local propagation, would you comment on this work with what you have found?

**Dr. Nelsen:** As I hinted at in the maximally stimulated case, which is 2 hr after progesterone or 8 to 10 hr after chorionic gonadotropin induction of ovulation, the majority of the waves seem to propagate a short distance. However, if you use the triple implant and analyze from gauge one to gauge three, in other words, about a 2 cm span, the number of waves propagated is much fewer than between 1 and 2 or 2 and 3 (1 cm spacing). So clearly, the propagation of waves over the length of the organ, such as occurs in the stomach or the small intestine, is rare, if it occurs at all.

**Dr. Neuman:** Do you think this technique could be used to separate longitudinal and circular muscle activity in various viscera?

**Dr. Nelsen:** Yes, we have done implants with adjacent longitudinal and transverse implants and they are easily separated.

Chapter 9

# LASER DOPPLER VELOCIMETRY USING FIBER OPTICS WITH APPLICATIONS TO TRANSPARENT FLOW AND CARDIOVASCULAR CIRCULATION

## Toyoichi Tanaka

## TABLE OF CONTENTS

# ABSTRACT

We present a new method of measuring the speed of fluid using laser Doppler velocimetry and a fiber optics. A 0.5-mm diameter plastic fiber optics inserted in the flow
transmits both the incident light and collects the light scattered by the moving particles
in the fluid. The spectrum of heterodyne beat between the scattered light and the local
oscillator, which originates at the end of the fiber optics, are measured using photon
correlation spectroscopy. We applied this technique to determining the velocity profile
of a transparent flow in a glass tube and the average blood flow velocity both in a
glass tube and in the femoral vein of a rabbit.

# I. INTRODUCTION

We should like to present a method with which to determine fluid flow velocities
using laser Doppler velocimetry and a fiber optics. The velocity of a flowing fluid can
be measured by determining the Doppler shift in the frequency of incident monochromatic laser light which is scattered from particles flowing with the fluid. Small frequency shifts associated with low-speed flow are determined using heterodyne beating
spectroscopy.[1] In this scheme, the frequency-shifted scattered light is combined on the
surface of a photomultiplier tube with a coherent local oscillator having the same frequency as the incident light. The beat notes between the scattered light and the local
oscillator appear in the photocurrent output of the photomultiplier and permit the
determination of the Doppler shift.

In this present paper, we report that flexible, small-diameter fiber optics can be used
to introduce the light beam into regions normally inaccessible to the light. In our experiments, the fiber optics not only brings the incident light into the flow region, but
also acts to transmit both the scattered light and a phase-coherent local oscillator produced at the end of the fiber optics out from the scattering region onto a photomultiplier surface.[2]

We have successfully applied this method to a study of the velocity profile of water
flow and of the blood flow velocities in both glass tubes and the femoral vein of a
rabbit (in vivo). We believe that this technique will prove to be a useful clinical technique for the study of blood flow in the cardiovascular circulation. It may also prove
to be a useful means for detailed measurements of moving objects in regions normally
inaccessible to light. It will also have useful applications in chemical engineering for
the monitoring of the flow of fluids and in mechanical systems for the study of moving
mechanical parts normally closed to visual inspection.

# II. PRINCIPLE OF LASER DOPPLER VELOCIMETRY

In this section, we explain briefly the principle of laser Doppler velocimetry (LDV).[1]
Let us consider a particle moving with a velocity, $v$, with a flowing fluid. We illuminate
the particle with a monochromatic laser light with frequency, $f_0$ (Figure 1). Due to the
Doppler effect, the scattered light has a frequency, $f$, different from the original frequency, $f_0$. The difference, which is called the Doppler shift, is proportional to the
particle velocity and is given by:

$$\Delta f \equiv f - f_0 \qquad (1)$$

$$\approx 2v \cdot \cos\theta/\lambda \qquad (2)$$

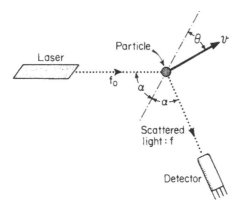

FIGURE 1. Schematic diagram of the scattering of laser light by a moving particle.

where $\Theta$ is the scattering angle as defined in Figure 1, and $\lambda$ is the wavelength of the laser light in the fluid. For $v = 1$ cm/sec and $\Theta = 60°$, for example, $\Delta f = 20$ kHz and $f_o/\Delta f = 3 \cdot 10^{10}$. To detect such a small change in frequency, we produce a beat of electric fields by mixing the scattered light having a frequency, f, with a local oscillator, which is part of the incident laser light having a frequency, $f_o$. The mixing takes place on the photomultiplier (PMT) surface. The beat appears as a sinusoidal oscillation in the PMT output $\Delta I(t)$ as a function of time, t:

$$\Delta I(t) = E_0 E_\varrho \cdot a \cos(2\pi\Delta ft) \tag{3}$$

where $E_0$ is the electric field of the incident light at the scattering point, $E_l$ is the electric field of the local oscillator on the PMT surface, and "a" is the attenuation of the scattered light field before it reaches the PMT surface. The PMT output is recorded in the form of correlation function of the photocurrent.

$$C_v(t) = <\Delta I(t + t')\Delta I(t')> \tag{4}$$

$$= E_0^2 E_\varrho^2 a^2 \cos(2\pi\Delta ft) \tag{5}$$

where $< >$ denotes the time average over $t'$. Thus, by measuring the frequency of the correlation function of the PMT photocurrent, we can determine the velocity of the moving particle.

If the scattering elements consist of particles having different velocities, the total correlation function C(t) is the sum of the correlation function $C_v(t)$ of photocurrent of light scattered from each particle:

$$C(t) = \sum_v C_v(t) \tag{6}$$

## III. EXPERIMENTAL ARRANGEMENTS

The experimental setup of the measurement of the blood flow velocity is shown in Figure 2. The laser light which is polarized perpendicular to the paper is divided by the beam splitter, BS. Half the initial light passes the beam splitter and is then trapped by the trap. The other reflected half goes into the lens, L1, and is focused onto the

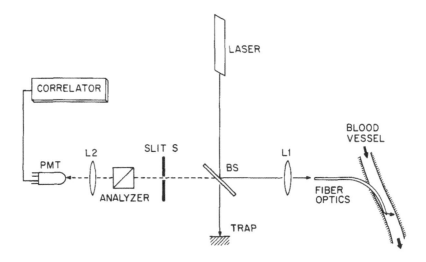

FIGURE 2.    Experimental setup of laser Doppler velocimetry using a fiber optics.

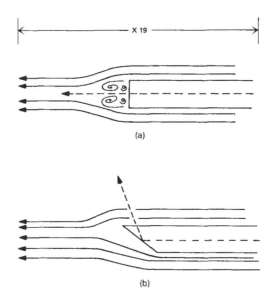

FIGURE 3.    Blood flow patterns around the fiber optics for two different types of exits. The dotted lines
show the laser light beams.

entrance of the fiber optics. This entrance is polished perpendicularly to the axis of
the fiber optics. The light passes through the fiber optics and goes into the flow. The
exit of the fiber optics is cut and polished in angles of either 90° or 30° (Figure 3).
The 90°-cut fiber optics was used for transparent flows since the light can reach far
into the laminar flow region. For a turbid flow, such as blood flow, we used the 30°-
cut fiber optics. In this case, the light is totally reflected into the side wall of the fiber
optics and illuminates the laminar flow region. In both cases, the light goes into the
flow and is scattered by the flowing particles. The backscattered light then comes again
into the fiber optic exit. It is partially reflected at the exit face, and the remainder goes
back through the same fiber optics to its entrance.

The laser light is also scattered elastically from the exit of the fiber optics. This light has the original frequency, $f_o$, and plays the role of local oscillator. Both the local oscillator and backscattered light are randomly polarized after passing back through the fiber optics. The backscattered light and local oscillator are then made parallel by the lens L1. Scattered light which comes into the exit with a large angle from the direction of the exact backscattering and, therefore, goes out at a large angle from the fiber axis out of the entrance is excluded by the slit S (see Figure 2). Thus, we can collect the light backscattered into a small solid angle around the initial light beam axis.

On emerging from the fiber entrance, the light scattered both from the particles and the fiber exit goes into the PMT, where they are mixed to give a beat note. The light reflected from the fiber entrance and the lens L1 could also go into the PMT, which could produce a high shot noise level. This light, however, is perpendicularly polarized and is completely excluded by the polarizer which is set parallel to the paper. However, half of the signal and the local oscillator light can pass the polarizer because of its random polarization after passing through the fiber optics.

The photocurrent output of PMT, which is proportional to the beat note of the scattered light, is then analyzed using the 18-channel digital autocorrelator.

## IV. PROFILE OF A TRANSPARENT FLOW IN A TUBE

We now apply the method to determining the profile of water flow in a glass tube. As scattering elements, we added one drop of 10% polystyrene sphere (diameter 0.091 μm) solution in 1 gal of water. The fiber optics was supported by a narrow (outer diameter 1 mm) glass tube and placed in parallel to the flow. Although fiber optics disturbs the flow, since the light can pass through water very far the observed scattered light comes mainly from the laminar flow region. Figure 4 shows the typical correlation function of photocurrent of scattered light. We observed an oscillating correlation function, whose frequency gives us the Doppler shift frequency and, therefore, the local velocity of the flow. In Figure 5, we show the measured velocity profile of a flow in a glass tube of diameter 6 mm at various radial positions in the tube. The measured data are plotted as solid circles. The solid line shows the parabolic profile whose absolute value was calculated using the flow output of the pump which produced the flow. The results show a very good agreement, and this justifies our belief that the fiber optics picks up the light scattered from regions well beyond its end.

## V. BLOOD FLOW IN A GLASS TUBE

Now that we have demonstrated that the method can be applied successfully to studying transparent flows, let us now examine flow of turbid fluid, blood. As was mentioned before, we used the 30°-cut fiber optics.

The reasons we used a fiber optics with sharply cut exit follow. In such a turbid system as blood, the light cannot penetrate far because of absorption and multiple scattering of light. If we use a fiber with a normal cut end, the light illuminates only a turbulent region at the head of the fiber optics (Figure 3). Therefore, we reflect the light to the side wall of the fiber, where the light can see the laminar flow.

Another reason is that in the in vivo measurements (next section) we do not have complete control of the position of the fiber optic end, i.e., the exit end is free. However, since the fiber is elastic and flexible at the same time, in most cases its exit lies on the wall of the vessel. In this position of the fiber, if we use a normal cut exit, the large part of the laser light would illuminate the vessel wall (see row C of Figure 6).

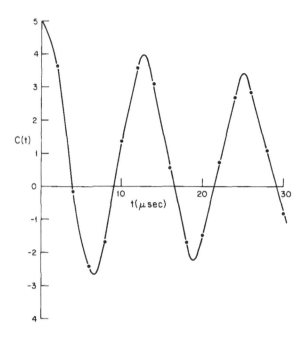

FIGURE 4.    The correlation function of the photocurrent of light scattered from polystyrene spheres moving with water flow in a glass tube.

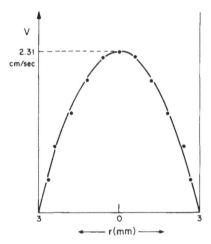

FIGURE 5.    The radial position dependence of the water flow in a glass tube with a diameter 6 mm measured by the present method. The solid line is the parabola calculated using the pump output.

On the other hand, in the case of sharply cut fiber optics, the light can illuminate the laminar flow region if the reflected light goes toward the center of the vessel (see row A of Figure 6). If the light strikes the wall of the vessel, the scattered light has no Doppler shift. In our in vivo measurements on rabbits, we first found this position where the correlation of the scattered light does not depend on time (See Equation 5), and then we rotated the fiber optics an angle of 180° to get the optimal position.

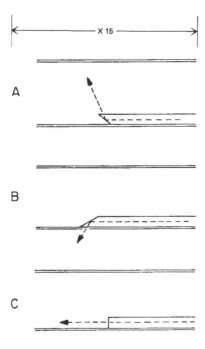

FIGURE 6. The path of the laser light emitting from the exit of fiber optics.

Indeed, this procedure was frequently used in our in vivo measurements and solved the problem of positioning the fiber optic catheter. In this optical arrangement, the light is scattered by erythrocytes moving with different velocities. Let us now calculate the total correlation function of light scattered from these erythrocytes.[2]

For simplicity, we consider the case where the fiber optic diameter is much smaller than that of the vessel. The initial laser light reflected by the exit of the fiber optics passes across the center of the vessel to the opposite side of the vessel wall.

We choose the exit end of the fiber optics as the origin of the coordinate, the diameter of the vessel as the z axis, and the direction of the blood flow as the x axis (Figure 7). The incident light intensity $E_0^2$ attenuates as the light passes through the blood and may be approximated by an exponential decay:

$$E_0^2(\vec{r}) = E_0^2(\vec{O})\exp(-r/\ell_0)$$

$$= E_0^2(\vec{O})\exp(-z/\ell_0\sin\theta) \tag{7}$$

where $\ell_0$ is the inverse of the turbidity of blood and is 0.33 mm (see Reference 3).

Using the same reasoning and the definition of a(r) we conclude that

$$a^2(\vec{r}) = \exp(-z/\ell_0\sin\theta) \tag{8}$$

The velocity distribution in the vessel may be approximated by the parabolic profile:

$$v(\vec{r}) = 2V_{av}(1 - [R - z]^2/R^2) \tag{9}$$

where R is the radius of the vessel, and $V_{av}$ is the average velocity of the flow, i.e., $\pi R^2 \cdot V_{av}$ is the volume of blood which passes through the cross section of the vessel.

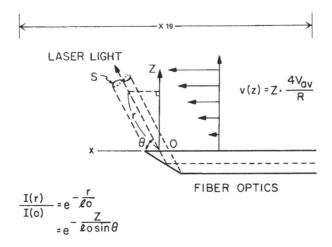

$$\frac{I(r)}{I(o)} = e^{-\frac{r}{\ell_o}}$$
$$= e^{-\frac{Z}{\ell_o \sin\theta}}$$

FIGURE 7.    The flow pattern around the fiber optics.

Using these equations we can now calculate the total correlation function:

$$C(t) = A \int_0^R \exp(-\alpha z)\cos[\beta v(z)t]\,dz \tag{10}$$

where $\alpha = 2/\ell_o \sin\theta$, $\beta = 4\pi \cos\theta/\lambda$, and $A$ = constant. Since the factor, $\exp(-\alpha z)$, decays very quickly as z becomes larger, the light scattered by the region of small z mainly contributes to this integration. In this region, the velocity can be approximated from Equation 9 as:

$$v(z) = 4V_{av}z/R \tag{11}$$

and the integration becomes:

$$C(t) = A \int_0^\infty \exp(-\alpha z)\cos(\gamma zt)dz$$

where:

$$\gamma = (4V_{av}\beta)/R = (16\pi \cos\theta V_{av})/\lambda R$$

On carrying out the integration, we find the Lorentzian correlation function:

$$C(t) = (A\alpha)/(\alpha^2 + \gamma^2 t^2) \tag{12}$$

Equation 12 shows that the correlation function is Lorentzian and that the half-width $\tau$ of the function, i.e., the value of time for which $C(\tau) = 1/2C(0)$, is given by:

$$1/\tau = \gamma/\alpha = [(4\pi \sin 2\theta)/R](\ell_o/\lambda)(V_{av}) \tag{13}$$

Thus $\tau^{-1}$ is proportional to the average velocity $V_{av}$ of the flow. The coefficient of proportionality is inversely proportional to the radius of the tube. $1/\tau$ is also proportional to $\sin 2\theta$, where $\theta$ is the angle between the incident laser light and the flow direction and $\ell_o$ is the penetration depth. Thus, we can expect $1/\tau$ to depend on the hematocrit. This analysis is, of course, oversimplified because we did not take into account the effect of multiple scattering and the effect of the size of the fiber optics

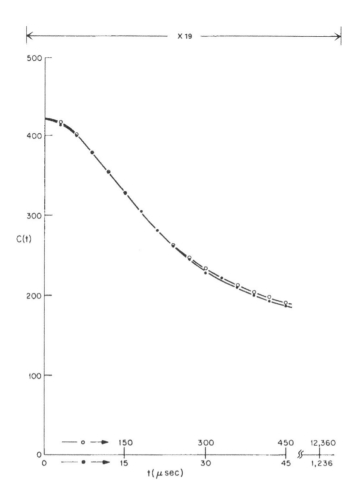

FIGURE 8.  Correlation functions of light scattered by the blood flowing in a glass tube with a diameter of 6 mm. The solid circles are for the
flow with an average velocity of 1.8 cm/sec, and the open circles are for
0.18 cm/sec.

in distorting the parabolic flow. These effects, however, should not affect the proportionality between $1/\tau$ and $V_{av}$, though it could affect the precise magnitude of the
coefficient of proportionality.

To check the formula (Equation 13) we measured the correlation function of the
scattered light for blood flow with known velocities in glass tubes. All the measurements were made using a fiber optics with a 0.5-mm diameter. The fiber optics is laid
on the wall of the tube in such a way that the light is reflected into the center of the
tube. We used fresh calf blood which was mixed with heparin. The glass tube, the
fiber optic catheter, and the reservoir of the blood were all siliconized to avoid the
coagulation of the blood on these walls.

Figure 8 shows the typical correlation functions of light scattered from blood flows
with different average velocities, $V_{av}$ (1.8 cm/sec and 0.18 cm/sec) in a 6-mm diameter
glass tube. The time scale of the two curves are scaled, i.e., the time intervals are 3
$\mu$sec for $V_{av} = 1.8$ cm/sec and 30 $\mu$sec for $V_{av} = 0.18$ cm/sec. We can see that their
shapes are almost the same, and that the inverse of the correlation width $\tau^{-1}$ is proportional to the average velocity of the flow. We should note, however, that the correlation curves have long tails and are not completely Lorentzian. Presumably, the multi-

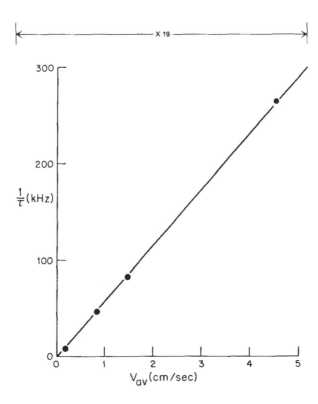

FIGURE 9.    The inverse of the correlation width of the scattered
light from flowing blood vs. the average flow velocity given from
the pumping rate to cause the flow.

ple scattering and finite fiber optic diameter effects mentioned above are responsible
for this.

Figure 9 is a plot of $1/\tau$ as a function of the average velocity, $V_{av}$, for the flows in a
tube with a 3.5-mm diameter. This shows clearly the linear dependence of $1/\tau$ on $V_{av}$
and demonstrates the validity of this proportionality as predicted in Equation 13. From
these measurements, we can conclude that for the flows in a tube with a given diameter,
we can obtain the average velocity by measuring the half-width of a correlation function of light scattered by erythrocytes.

## VI. IN VIVO MEASUREMENTS OF BLOOD FLOW SPEED IN A FEMORAL VEIN OF A RABBIT

So far, we have discussed the blood flow in glass tubes. Let us now consider the
blood flow in vivo. Using the present method, we measured the blood flow speed in
the femoral vein of a healthy albino rabbit whose weight was 5 kg. The vein had a 2-
mm diameter. We exposed the vein and inserted the fiber optic catheter through a
small hole made by a needle. The catheter was inserted about 8 cm from the hole.

First, we found the position of the fiber optics, where the light struck the vessel
wall, and we observed no Doppler shift. Then we rotated the fiber optics by 180° to
obtain the optimal position as was shown in Figure 5. Figure 10 shows the correlation
function of the scattered light. The half-width was 14 μsec, which corresponds to the
average velocity of 2.1 cm/sec.

At the end of the experiment, the rabbit was killed by adding enough anesthesia

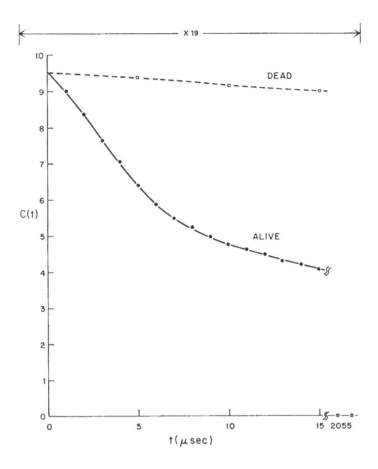

FIGURE 10. The correlation functions of the scattered light from the blood flow in the femoral vein of an albino rabbit. The lower curve was measured before the rabbit was killed, and the higher one was measured after its death.

through a needle which had been set in advance in the artery of the ear of the rabbit. The correlation function then became almost flat, which indicates that the velocity of the blood flow became zero (Figure 10).

As far as we know, there are no data with which to compare our results. Since neither modern technique is relevant for such a small vein or such a slow speed, we employed the most primitive method. We cut the femoral vein and accumulated the blood while it was flowing out. The flow rate was 3.0 m$\ell$/50 sec, which corresponds to $V_{av}$ = 1.9 cm/sec and agrees with the data we obtained using laser Doppler velocimetry. For this experiment we used a different, but similar, rabbit.

## VII. CONCLUSION

Using laser Doppler velocimetry and a fiber optics, we were able to determine the velocity profile of a transparent flow and the average velocity of blood flow in the femoral vein of a rabbit. The apparatus is capable of detecting the values of the average velocity ranging from 0.01 to 100 cm/sec.

We conclude that laser Doppler velocimetry using a fiber optic catheter may provide a quick, simple, and accurate means to find average blood flow velocity, particularly in small-diameter vessels having low blood velocity. It would also be applicable to a wide variety of applications in chemical and mechanical engineering.

## ACKNOWLEDGMENT

The author should like to thank Professor G. B. Benedek for helpful discussions and suggestions. This work is supported in part by NIH, Grant EY 02433 and NSF, Grant CHE 77-26924.

## REFERENCES

1. Chu, B., *Laser Light Scattering,* Academic Press, New York, 1974.
2. Tanaka, T. and Benedek, G. B., *Appl. Opt.,* 14, 189, 1975.
3. Anderson, N. and Sekelj, P., *Phys. Med. Biol.,* 12, 173, 1967.

## DISCUSSION

**Dr. Sidney Wolfson:** Since, as you mentioned, the intensity of the light beam and the scattered light falls off exponentially as it penetrates the flowing blood, then there should be some limit where the returning light would be negligible, and I don't understand the term that involves the radius of the vessel in a very large vessel, unless that applies to some other aspect of the calculation.

**Dr. Tanaka:** It is because we measure the velocity gradient of blood flow near the vessel wall. The velocity gradient is inversely proportional to the radius of a vessel for the same average velocity.

**Dr. Wolfson:** You are correcting for the flow profile in the blood vessel and assume that the fiber optic is near one edge.

**Dr. Tanaka:** That is right.

**Dr. Wolfson:** Is there any sort of provision by which this could be used to measure volume flow under certain circumstances?

**Dr. Tanaka:** In this technique, we must know the radius of the vessel to determine the average velocity. That means that we should know both the radius and the average velocity. Then, assuming a parabolic profile, we can determine the volume flow.

**Dr. Wolfson:** How will you estimate the radius in a deep lying vessel?

**Dr. Tanaka:** We do not have a good idea. Please think of a good method.

**Dr. Richard S. C. Cobbold:** My question concerns the assumptions of your mathematical model. It seemed to me that you are assuming that the light is transmitted parallel to the axis of the fiber and that you were, in effect, dealing with a single angle of emission. Of course, one has multiple reflections, and it seemed to me that at that angle the cone at which the light emerges would indeed be quite broad and that that would have quite a profound influence on your results.

**Dr. Tanaka:** There is a spreading of the laser beam after emerging from the fiber optics. The spreading can explain the broadening of the Doppler shift. However, the angle of spreading is within 10° and is not so large as to smear out the Doppler shifted peak. Therefore, we can determine the absolute value of the local velocity from the Doppler shift frequency.

**Dr. Isaac Greber:** With regard to the previous discussion, I think there is a much more significant error in the assumption that the procedure depends very strongly on the ability to calibrate. Probably the strongest disturbance is the fact that you're putting the probe into a region in which the velocity is small. If you go ahead now and consider that you place a probe and it makes a small disturbance on the linear profile, then, in fact, one can estimate the effect of this disturbance. If you disturb this linear profile in a way that's predictable, you stay clear of that disturbance by the calibration procedure.

**Dr. Tanaka:** Yes, I agree.

**Dr. Walter Olson:** At the probe, you have a single bevel or angle, and you turn it to 180° to assure your positioning. Did you consider taking two paths simultaneously to monitor positioning at all times? In other words, you could do them both at once.

**Dr. Tanaka:** Yes, that's right, it is possible.

**Dr. Peter W. Cheung:** I have several specific questions. What is the angle of the fiber's bevel?

**Dr. Tanaka:** Approximately 30°. Then the edge of the fiber optics reflects light totally.

**Dr. Cheung:** Do you know the extent of the penetrating depth of the light into the blood at normal hematocrits?

**Dr. Tanaka:** Less than 0.3 mm.

**Dr. Cheung:** Is that a measurement or a calculation?

**Dr. Tanaka:** It is a measurement for calf's blood.

**Dr. Cheung:** This is a direct transmission measurement?

**Dr. Tanaka:** Yes, we passed a straight beam into a small amount of blood and measured the transmission at a far distance. By changing the optical path length we could determine the turbidity and penetration depth of the blood.

**Dr. Cheung:** Approximately how much laser power do you need, and what is the signal-to-noise ratio of your overall system?

**Dr. Tanaka:** The power we used is one mW in the helium-neon laser. The signal-to-noise ratio was so good that we were not concerned about it.

**Dr. Sidney Wolfson:** I have thought of a way to measure the radius. It'll involve making this a little bit thicker and putting a little more hardware in there which might make it less suitable. Since you require rotation and orientation, if you put a piezoelectric crystal and orient properly with your angle, then you can use an echo to measure the diameter.

**Dr. Charles F. Knapp:** On the parabolic profile that you show in water, how do you know how far above the fiber you are actually measuring? Where is your focal point or your sample volume?

**Dr. Tanaka:** The scattered light measured came from a region within about 1 cm from the fiber optics end.

**Dr. Knapp:** You take a measurement, you get a reading, then you move it up to a new radius position. How do you know that you haven't taken an ensemble average all the way across the vessel?

**Dr. Tanaka:** Because the angle of spreading is not so large. Also, the center of that cone of the light beam is the largest in intensity. We are measuring mainly the light scattered backward along the cone axis.

**Dr. Knapp:** Your estimate on the distance is how much?

**Dr. Tanaka:** 1 cm.

**Dr. Knapp:** You're taking values 1 cm from the side wall of the fiber optic?

**Dr. Tanaka:** No, off the end of the fiber.

**Anon.:** Can you use a balloon to measure the radius?

**Dr. Tanaka:** You might use a balloon and make it swell to measure the radius and also control the position of the fiber in this way.

**Setsuo Takatani:** When you get a parabolic profile, what did you use as a fluid?

**Dr. Tanaka:** We used distilled water having a slight amount of polystyrene spheres. If you measure water, especially Cambridge water, it has a lot of dust in it, so you don't have to add anything to it.

**Maurice Karkar:** Can you comment on the effect of cell shape and cell size on your measurements?

**Dr. Tanaka:** We haven't done any experiments on the effects of size and shape. I can imagine that the cell shape or size might alter the flow profile and produce a difference in measured correlation function. Also, if the cells have a nonspherical shape, then they can rotate, and this rotational movement can cause a Doppler shift. However, since this effect is very, very small compared to the translational motion, it is not significant. The largest effect would be the change in flow profile in the vessel.

**Dr. Peter W. Cheung:** If I understand correctly then, the penetration is on the order of less than 1/2 mm, and when your fiber is in a vessel and you cut it at an angle, in coming out you are measuring velocity which is approximately 1/2 mm away from the fiber, but you can't measure anything further than that.

**Dr. Tanaka:** We are measuring the gradient of the velocity.

**Dr. Cheung:** How do you have assurance that the fiber is staying at the side of the wall or any other place in the vessel?

**Dr. Tanaka:** You need experience. In this kind of fiber when introduced to the vessel, it is always lying on the vessel wall because it's elastic. It would like to be straight and the only way to satisfy that condition is lying on the vessel wall.

**Dr. Michael R. Neuman:** Could this technique be used to measure something like cilliary beating?

**Dr. Tanaka:** Oh yes, I think that has been done.

Chapter 10

# OBTAINING PROPRIOCEPTIVE INFORMATION FROM NATURAL LIMBS: IMPLANTABLE TRANSDUCERS VS. SOMATOSENSORY NEURON RECORDINGS

Gerald E. Loeb, Bruce Walmsley, and Jacques Duysens

## TABLE OF CONTENTS

# I. INTRODUCTION

The paralyzed limbs of quadriplegic and paraplegic patients constitute one of the most intriguing challenges for applied neural control techniques. Each essentially intact limb contains the results of millions of years of evolutionary progress: efficient chemical to mechanical energy conversion, temperature regulation, self-lubricating joints, high-gain amplification of control signals, sensitive detectors of mechanical events and feedback, and a wear-resistant contour enclosure. Even large parts of the feedback circuits for regulating both external actions (e.g., muscle stretch reflexes) and internal conditions (e.g., vascular reflexes) usually exist intact in the spinal cord. The system lacks only the descending control signals from the brain and a means of reporting its activities back to conscious levels.

It has now become possible to utilize at least the mechanical output system in these patients by functional neuromuscular stimulation (FNS). Electrodes are implanted in muscles to stimulate motor neuron terminations on muscle fibers, restoring functional contraction to the muscles. The patient can use his own limb to perform tasks such as grasping utensils and raising food to his mouth.[1] The two major limitations hindering extension of such techniques to more sophisticated functions such as walking are (1) obtaining multichannel command and control signals and (2) obtaining sensory feedback for both conscious sensation and unconscious regulation of stimulator output. While the former has been the subject of considerable study,[2] the need for the latter is just becoming evident and will continue to grow as the devices address more complex tasks.

There are three general classes of techniques which are potential candidates for obtaining such somatosensory information from the FNS limb. Most obviously, a brace or glovelike structure containing conventional electromechanical transducers can be applied externally to the limb as an orthotic device. Such a device is relatively easy to design, modify, and service, but suffers from the fatal flaw of being psychologically and cosmetically unacceptable to most patients.[3] In addition, such devices cannot avoid interfering with and limiting free use of the limb. Two more acceptable techniques, implanted transducers and implanted electrodes recording naturally occurring mechanoreceptor activity (via their afferent nerve fibers), can be compared by their potential for long-term survival in the hostile environment of the body.

Any implanted device must contend with three general complications brought on by this environment. First, the device is essentially isolated from monitoring, calibration, regular servicing, or simple repairs. This means guaranteeing both normal functioning and complete freedom from tissue-damaging failure modes for at least a decade. Second, the device is essentially submerged in warm salt water containing small concentrations of lipids, proteins, and digestive enzymes. No material is known to be totally impermeable and insoluble to this combination, particularly with the added constraint of minimizing the encapsulating and inflamatory responses evoked in the surrounding tissue. Finally, the body is constantly moving and remodeling its structure. No internal site is completely free from this physical plasticity, which the naturally occurring body parts survive only by themselves constantly remodeling and renewing their parts and surfaces.

The general class of implantable transducers will here be represented by two devices which we have constructed for physiological research, a strain gauge which clips onto intact tendons and a length gauge which is stretched between two rigid tethering points. As summarized in Appendix 1, such devices for short-term (1 to 3 month) chronic use are fairly easy to develop and implant, and they give direct indications of the variable being detected. However, it should be noted that for certain detection tasks of interest (e.g., local skin pressure, stretch, and slip), no satisfactory implantable devices yet exist.

For long-term applications such as prostheses, the suitability of all such devices is uncertain. Experience to date indicates that implantable electronic circuits involving active elements, critical component values, and/or applied voltages can survive indefinitely only if hermetically sealed in rigid ceramic or metal cans.[4] Such encapsulation techniques are, of course, not feasible for a device which must be directly influenced by mechanical events surrounding it. Experience also indicates that it is very difficult to maintain mechanical linkages to living tissue under constant stress and movement because of the tendency for even small forces to restrict local blood supply or stimulate erosion and remodeling of the tissue.

Since the FNS patient already has a highly sophisticated array of force, length, and pressure transducers in his limb, i.e., the somatosensory mechanoreceptor neurons, it should be possible to avoid many of these engineering problems. Recording the activity of these neurons under acute dissection conditions has been well understood, but generally has involved techniques not amenable to chronic implantation (e.g., dissecting nerve filaments in oil pools, inserting rigid, sharpened microelectrodes into mechanically immobilized nerves). There have been attempts to place whole nerve bundles in insulating cuffs which, theoretically, should amplify the tiny action potential currents associated with each nerve fiber by restricting the conductive extracellular space.[5] However, in practice, these cuffs have had to be large to avoid damaging the nerve or its blood supply and have usually produced signals representing the sum of large numbers of very small signals from heterogeneous signal sources.[6]

The concept of floating microelectrode arrays has changed these limitations, but substituted others. Typically, 10 to 20 microelectrodes with ultraflexible lead wires are inserted individually into target structures such as brain stem,[7] cerebral cortex,[8] spinal roots,[9] or spinal ganglia.[10] Each tip records unitary extracellular action potentials from one or a small number of adjacent neurons (usually cell bodies). Because of the small electrode tips and flexible lead wires, the activity of a single neuron can be reliably distinguished for up to several weeks, but there is no control over the apparent slow, random drifting of the electrode with respect to the neurons. Although unit activity continues to be obtainable for over a year,[11] the particular information being derived is slowly and uncontrollably drifting.

For a prosthesis to make reliable use of this information, it would be necessary to have a large number of such microelectrodes connected to an intelligent central processor unit which would keep track of the units' waveforms and recognize the loss or addition of new units. Such units would be identified by their correlation with existing units and then become part of the control ensemble, a process not unlike that suggested for the developing nervous system.[12] Recent and continuing advances in microelectronics make such an intelligent processor quite achievable within realistic constraints on space and power. In fact, the major limitation on implantable electronics appears to be biocompatible, hermetic packaging and feed-through of leads, a problem which is essentially unchanged by the complexity of the device's logic.

The main advantage of this technique for obtaining somatosensory information lies in the extreme simplicity of the most vulnerable part of the electronics, the tissue interface. The microelectrodes can be located in the most stable part of the trunk, the spinal canal, and still draw on information conducted there by fibers from the most distal extremities. The microelectrode tip is a passive, inert detector of normally present weak electrical fields, requires only a modest insulation layer, and is insensitive to water vapor permeation. Although the initial surgical implantation involves some tissue trauma and nerve cell loss, most neural systems have very great redundancy and can remove debris readily. In fact, it appears that such local tissue remodeling is responsible for the remarkably clean, stable potentials obtainable chronically with relatively crude, large electrode tips. The types of information generated by the various modalities of receptors during normal walking are discussed below.

## II. IMPLANTABLE TENDON STRAIN GAUGE

Perhaps the most obviously needed transducer in an FNS system is that for the force generated by the stimulated muscles. Such information can be used to linearize and regulate the stimulation-generating algorithm and may be usefully displayed to the patient as an indication of forces being exerted on external objects.

Figure 1 shows the design and application of such a gauge which can be clipped onto the intact tendon in a minor surgical procedure.[13] The active element is a commercially available semiconductor strain gauge (BLH type SPB1-12-35) which is bonded with epoxy to a stainless steel substrate. Flexible stranded stainless steel wires connect the device to a DC bridge circuit which detects resistance changes resulting from substrate deformations induced by tendon strain. The stainless steel substrate is machined in the shape of a rounded capital "E", which is clipped onto and effectively captures the individual tendon as indicated. The device is electrically insulated with a conformal coating ($15\mu$m) of Parylene-C®[14] and mechanically protected with a thin layer of Silastic® silicone rubber (Dow Corning® Medical Adhesive A). With appropriate sizing of the aperture, such gauges have been left attached to the individual components of the Achilles tendon of cats without visible erosion or other damage to the tendons and without interfering with normal movements over a 2-month period.

The particular sensitivity of a given gauge is determined by the mechanics of the stainless steel substrate. Linearity to within the resolution of the detecting circuitry has been achieved for passive external loads and active muscle contractions up to 12 kg. The gauges have excellent frequency response (17 kHz response to pulsed unloading), high sensitivity (350 Ω rest resistance with 5 Ω linear range), and excellent stability (calibration unchanged after 2 months in vivo).

Figure 2 illustrates the output of two such gauges implanted on the soleus and medial gastrocnemius muscle tendons in the Achilles tendon of a cat. The sequence of events represents jumping to reach an object along with preparatory and following activity. The force of the two muscles may be compared with the EMG potentials, which are seen to be considerably phase advanced during fast movements. The two muscles, which are dominated by slow-twitch and fast-twitch units, respectively, differ in their usage patterns. The "slow" postural soleus muscle reaches a similar peak contraction for all activities (although nonisometric tension rise time can be quite brief as a result of cross bridge dynamics), whereas the "fast" gastrocnemius muscle provides the range of forces required by the different activities.

## III. IMPLANTABLE LENGTH GAUGE

If the FNS patient is to benefit consciously or unconsciously from a kinaesthetic sense of limb position (probably necessary for tasks such as ambulation), then he must be provided with some indication of joint angles and/or muscle lengths. Figure 3 depicts the construction of a device which we have successfully implanted in multiple sites in the cat hindlimb to detect the angles between skeletal segments. Signals corresponding to the length of the stretched rubber tubing have been obtained for over 2 months in a freely moving animal without apparent interference with its normal activity.

FORCE TRANSDUCER

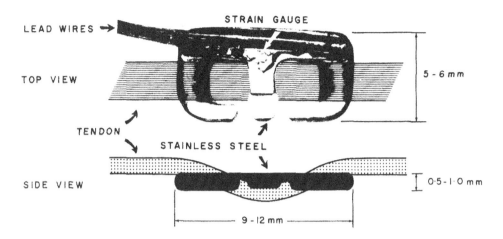

FIGURE 1. Photograph of the transducer assembly (*en face* view above) indicating placement of semi-conductor strain element lead wires and position of tendon (shading). Lower diagram shows longitudinal section with tendon (stippled) threaded through gauge.

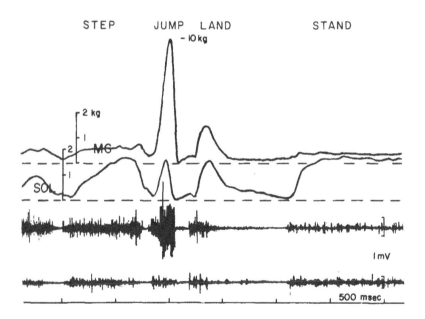

FIGURE 2. Records of MG and SOL force and EMG from a 2.9 Kg cat during a sequence of movements during preparation for ("step"), and landing after ("land-stand"), a vertical jump to a table 70 cm above the floor. "Stand" portion represents quadripedal standing during eating of food reward.

The principle is that of detecting the change in electrical resistance of the conductive fluid column which has a constant volume in an elastic tube. The device shown is essentially a modification of a similar design employing mercury as the conductive liquid and a frequency modulated oscillator as the detector.[15] The use of hypertonic saline gels instead of mercury confers several advantages. The device is considerably easier to construct and can be simplified even further than the version shown here.

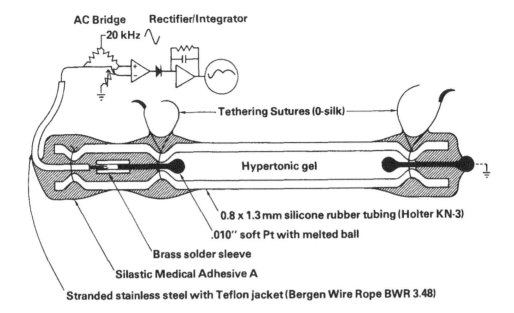

**FIGURE 3.** Schematic of the construction of the implantable length gauge and its associated electronic detecting circuitry. The tethering sutures are used both to seal the conductive fluid inside the distensible Silastic tubing and to surgically anchor the device to burr holes or screws in bony structures. The change in electrical resistance between the two electrodes (one grounded in the extracellular fluids of the limb) as the gap increases is sensed by an AC bridge and converted into a voltage proportional to the length of the stretched gauge.

(At high enough frequency, electrode surface impedance becomes negligible and the stainless steel wire can itself form the electrode, eliminating a solder joint). In case of catastrophic failure of the tubing, the contents are nontoxic. The decreased mass of the fluid increases the resonant frequency considerably beyond the 85 Hz linearity limit encountered with mercury (see Figure 4B). The silicone rubber tubing acts like a dialysis membrane with respect to the hypertonic contents, causing the gauge to absorb water and equilibrate with isotonic body fluids by becoming internally pressurized, decreasing the tendency for the tubing to be compressed or kinked by lateral pressure from adjacent tissues. Finally, the impedance range of the device is typically 10 to 20 kΩ. This allows the grounding of one end of the gauge directly to body fluids and using only a single fine-gauge lead wire instead of the two relatively large, low-resistance leads needed for the mercury device (typically less than 1 Ω). The main disadvantage of this technique is that gauges cannot be stored indefinitely in air because of evaporation of water through the silicone rubber and its replacement by inward diffusing air. We store them submerged in sterile isotonic saline, which also keeps them near equilibrium pressure. They are completely autoclavable if kept submerged.

Figure 5 shows such gauges implanted at three sites in the cat hindlimb from which the three traces of hip, knee, and ankle angle were obtained during walking. The hip gauge was tied proximally to a stainless steel bone screw set in the anterior iliac crest and passed s.c. to its sutured attachment to the periosteal connective tissue just proximal to the knee joint capsule. The knee gauge was tethered between a bone screw in the posterior mid-shaft of the femur and a hole drilled through the anterior tibial ridge, passing between the hamstring muscle bellies and s.c. medial to the tibia. The ankle gauge originated at the same tibial burr hole and passed s.c. medial to the triceps surae muscles to its attachment at a burr hole through the most posterior part of the calcaneus.

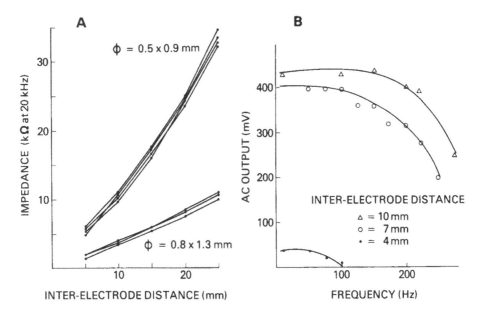

FIGURE 4. (A) Linearity and reproducibility of several gauges of different tubing sizes filled with 5X isotonic saline (inner × outer diameters given). (B) Frequency response to small-amplitude sinusoidal vibration applied to a length gauge at various degrees of stretch (4 mm is the slack point for the gauge).

The gauges were calibrated in angular degrees by picking two arbitrary points near the extremes of movement during the step cycle and determining the joint angles from videotape stills. The complete step cycle was analyzed for joint angles as indicated by the dots. Light spots from small reflective disks glued to the shaved skin at palpatable landmarks were used to construct vectors traced over the videotape image from which the angles were measured.

The rather considerable lack of agreement between the two methods in some parts of the step cycle reflects on the rather different errors to which each method is prone. The length gauge is, of course, not an angle gauge. A considerable error occurs in failing to correct for the cosine function relating the length of the hypotenuse being measured to the joint angle, particularly for the large excursions away from 90° experienced at the ankle. A small error also arises from the parabolic, rather than linear, relation of resistance to length, but this is small for the length changes involved (less than 20% of rest length). A less correctible error apparently occurs in the ankle gauge during stance because of medial bulging of the tensed calf muscles. This forces the gauge along a curved s.c. path between the tibial origin and calcaneus insertion (i.e., the gauge is correct for arc rather than cord muscle length). The videotape method is also prone to significant errors. The skin is known to slide with reference to the boney landmarks,[16] and the center of rotation of the joints may be changing (both particularly important around the knee). The limb does not move entirely parallel to the focal plane of the camera during stepping, but undergoes considerable internal rotation plus inversion and abduction. This produces 5 to 10° errors in the apparent angle seen at the ankle. In our experience, if a path for the gauge can be devised which is reasonably straight and in which it is not kinked by crossing muscles or folding skin, the gauge is an accurate measure of the length changes occurring between two points. This length may itself be of interest (e.g., directly related to a muscle action) or may require considerable calculation to convert to an externally apparent angle.

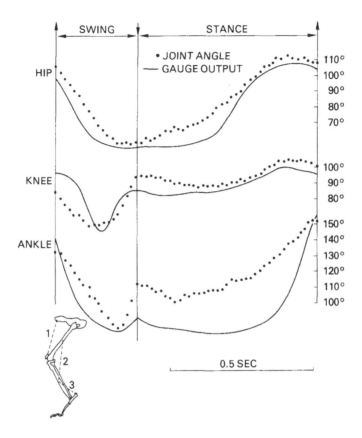

FIGURE 5.  Simultaneous records from three length gauges implanted
across the hip, knee, and ankle joints of a cat as shown in the detail at
the lower left. The dots indicate the joint angle measured from single
fields of a simultaneous videotape record of the movements.

## IV. CHRONIC SPINAL GANGLION ELECTRODE ARRAYS

Relatively stable, chronic recordings of single-unit somatosensory signals can be obtained from the primary afferent cell bodies located in the dorsal root (spinal) ganglia (DRG).[10] Although the site poses somewhat difficult surgical access (particularly at cervical levels) and is subject to some movement, the cytoarchitecture of the DRG is distinctly advantageous. The cell bodies are very large (40 to 80 μm diameters even for small diameter nerve fibers) and are densely packed together in a tough connective tissue matrix. The large extracellular potentials generated by action potentials invading the unipolar cell bodies are augmented and modified by the glomerular approaching axon and the capsule of glia-like satellite cells, typically resulting in two to four discriminable unit records per electrode. The dense connective tissue stroma anchors electrodes and stabilizes each unit for several days to weeks, even without resorting to surgical fixation.

Figure 6 shows the surgical approach and electrode design used for neurophysiological studies in the cat hindlimb. The "microelectrodes" are simply the cut ends of 50-μm-diameter insulated wires which are inserted by hand-held forceps into the L7 and S1 DRG through the intact dural-epineural sheath. The simple platinum-iridium alloy wires provide enough flexibility to take up the small sliding movements between the DRG and the fixation point of the silicone rubber carrier tube to the L7 dorsal spine.

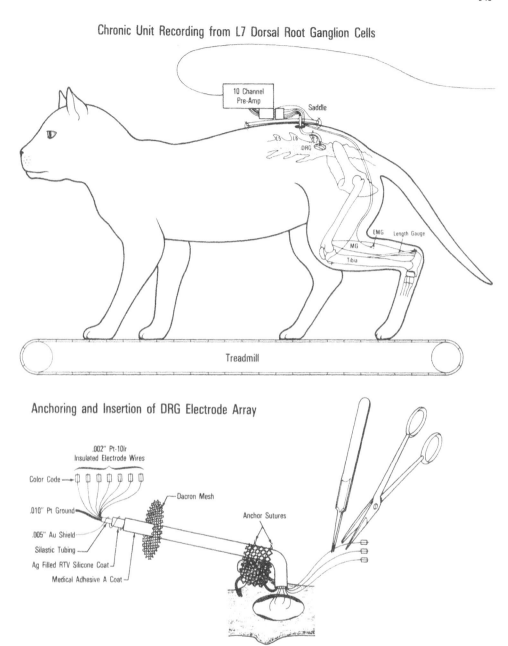

Chronic Unit Recording from L7 Dorsal Root Ganglion Cells

10 Channel Pre-Amp

Saddle

L5  L6

DRG

EMG  Length Gauge

MG

Tibia

Treadmill

Anchoring and Insertion of DRG Electrode Array

.002" Pt-10Ir
Insulated Electrode Wires

Color Code

Dacron Mesh

Anchor Sutures

.010" Pt Ground

.005" Au Shield

Silastic Tubing

Ag Filled RTV Silicone Coat

Medical Adhesive A Coat

Cephalad ◄── L7 Vertebra ──► Caudad

FIGURE 6.  Technique for implanting and recording from primary afferent neurons in the dorsal root ganglia of freely walking cats. The insulated microelectrode wires are cut obliquely to expose a sharp, bare tip which is inserted manually into the intact DRG. A loop of slack left between the DRG and the silicone rubber carrying tube takes up the movement of the DRG during walking. No other fixation is used until the wires terminate in a permanently mounted percutaneous saddle connector. (From Loeb, G. E., Bak, M. J., and Duysens, J., *Science*, 197, 1192, 1977. With permission.)

The obliquely cut wire ends provide relatively large electrode surface areas having tip impedances of 150 to 250 kΩ (at 1 kHz). This results in lower thermal noise and better stray signal immunity than conventional microelectrodes (typically, 1 to 10 mΩ tip impedances). The considerable tissue damage almost certainly resulting from the initial

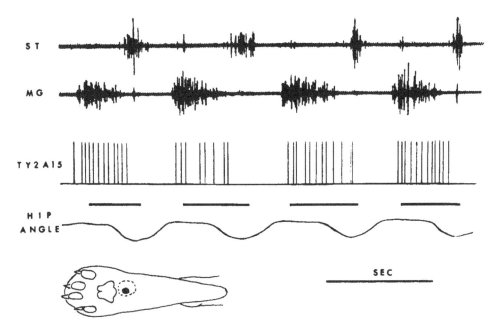

FIGURE 7.   Activity recorded from a slowly adapting type-II (stretch sensitive) light-touch receptor in the hairy skin of the cat hind paw (receptor field cross-hatched, and area of stretch sensitivity outlined in detail at lower left). The top two traces show EMG activity recorded during slow, regular walking (ST = semitendinosis muscle, MG = medial gastrocnemius muscle). The black bars indicate the stance phase of the gait (derived from simultaneous videotape analysis), and the bottom trace indicates hip swing (extension downward). The afferent unit activity (TY2A15) is shown as acceptance pulses from a spike discriminator set to respond to the waveform of this unit. It indicates activity correlated with the stretch of the receptor field skin just preceding footfall (E1 extension) and during dorsal yielding of the toes in stance (the receptor field is never in direct contact with the treadmill surface).

manual insertion of these large probes appears to be readily cleared in the first 24 hr. Typical action potentials are 100 to 400 $\mu$V peak to peak, with very distinctive and stable waveforms. The implants have been maintained for over 2 months with unit activity.

The entire range of myelinated (and some unmyelinated) fibers is sampled by this technique. Of 137 units identified in cat L7 and S1 DRG, approximately 40% were proprioceptors (predominantly muscle spindle primary and secondary endings plus a scattering of Golgi tendon organs and joint receptors), and 40% were cutaneous receptors (predominantly hair cell endings plus slowly adapting receptors of light pressure and skin stretch). The remaining 20% were spontaneously active, but insensitive to manipulation of skin or joints, leading us to suspect visceral or chemoreceptive functions.

The activity of the cutaneous receptors was quite stereotyped during regular walking and usually showed fine sensitivity to appropriate stimuli local to the receptor ending. The slowly adapting light-touch receptors would be anticipated to be of most interest to FNS engineers because of their demonstrated ability to respond with stable firing levels in proportion to skin deflections during acute physiological experiments. We have shown that these receptors continue to be reliable indicators of skin pressure and stretch during normal activity. SA-I class receptors are sensitive only to direct skin pressure, and their activity as recorded chronically generally followed the local skin pressure as best it could be estimated from available kinesiological data. Stretch-sensitive SA-II receptors were generally modulated or entirely activated by skin stretch or shear, as shown in Figure 7. We know of no implantable, or even externally applicable,

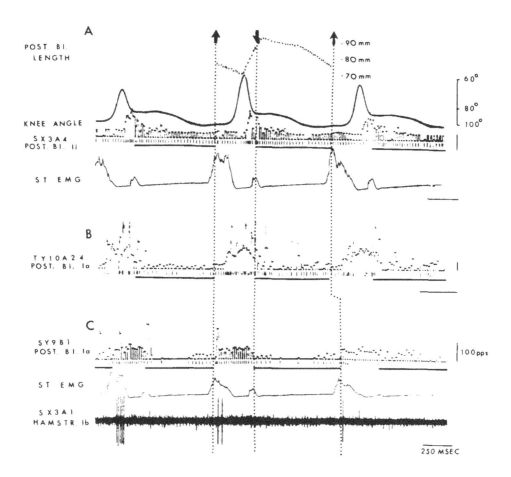

FIGURE 8. Three similar step cycles illustrating the behavior of various proprioceptive afferents from the hamstring muscles of the cat hindlimb. The top trace indicates approximate muscle length as obtained from videotape analysis of the step cycle in A, dotted vertical lines indicate alignment of foot lifts (up arrows) and foot fall (down arrow) for all three examples. Afferent activity is shown as both spike acceptances by the discriminator (vertical tick marks) and instantaneous spike frequency in pulses/sec. (A) Spindle secondary ending from posterior biceps muscle (43 m/sec conduction velocity) illustrating sensitivity to overall muscle length and lack of response to active state of muscle contraction (rectified and integrated EMG from the synergistic semitendinosis muscle at bottom). (B) and (C) Responses of two posterior biceps muscle spindle primaries (SY9B1 = 105 m/sec conduction velocity) showing lack of correlation with muscle length and probable influence from fusimotor contractions coactivated with the extrafusal muscle contraction during swing phase. Bottom trace shows the actual signal from a microelectrode recorded simultaneously with unit SY9B1 in which the only activity came from an afferent, later identified as a Golgi tendon organ, in one of the hamstring muscles.

transducer which could signal such events without distorting them.

The proprioceptors were somewhat more complex. The naturally occurring tendon strain gauge, the Golgi tendon organ, appeared to be an excellent monitor of active muscle tension as best it could be estimated from integrated EMG records (see Figure 8C). A direct comparison of tendon organ output with simultaneous tendon clip strain gauge signals remains to be obtained. The joint receptors were generally active during walking, but had activity which was not simply related to the extension/flexion angle of the knee (or its velocity), and probably were influenced by joint loading and inserting muscles, as suggested by others.[17]

The naturally occurring muscle length gauge, the muscle spindle, is complicated by its internal gain setting mechanism, the intrafusal muscle fibers (gamma and beta innervated motor units in series with the stretch-sensitive receptor endings). During regular walking, the spindle secondary endings were generally reliable indicators of true muscle length, indicating either zero or constant levels of fusimotor influence (Figure 8A). Spindle primary endings (and secondary endings during nonstereotyped behavior) generally exhibited variable and often complex activity modulations which could be attributed to both passive stretch properties and active fusimotor input (Figure 8B and C).

It is difficult to predict the nature of this fusimotor influence to be expected in the FNS patient. Although the descending influence on gamma and beta motoneurons has been lost along with the alpha motoneuron control, the local reflex loops are still vigorously active by clinical measures. Electrical stimulation of muscle nerve afferents and efferents in FNS may itself set off spinal reflex activation of fusimotor neurons to the stimulated and/or other unstimulated muscles. (In one cat experiment, an electrical stimulation of bipolar electrodes implanted in the gastrocnemius muscle, not unlike an FNS input, evoked a short latency fusimotor response in an antagonist muscle, peroneus longus, manifested by a burst of activity from a spindle primary in that muscle.) Chronic electrical stimulation of muscles has been reported to lead to changes in the spastic hyperreflexia experienced by many spinal injury patients.[18] These changes may be mediated by resetting of background fusimotor excitability. Just as the chronically stimulated muscle changes its contractile properties,[19] the receptor may change its sensitivity, requiring flexible and sophisticated processing of its signal to achieve useful muscle length feedback control.

## V. CONCLUSION

Although the naturally occurring transducers of length, strain, and pressure have obvious counterparts in electromechanical devices, they are not entirely characterizable by the usual engineering criteria. Being living tissue, they can adapt and change their properties over time and circumstances, a process not equivalent to engineering's "drift" factor because it is purposeful behavior in response to the environment and cycles of growth and development. Animal experimentation must be carefully tailored to make the circumstances as closely parallel to those expected in humans. Even then, each FNS patient will represent a unique sequence of temporal and physical factors which will influence both the afferent and efferent interactions with his prosthetic system.

# APPENDIX

## PROBLEMS ENCOUNTERED BY IMPLANTED DEVICES:

I. Changes in electro-mechanical characteristics from absorption and permeation of body fluids

II. Loss of electro-mechanical integrity from constant body movement

III. Inaccessibility for monitoring, calibration, service, and repair

### IMPLANTABLE TRANSDUCERS:

I. Muscle Tension - tendon clip strain gauge
II. Muscle Length - fluid-filled silicone tube

### TRADE-OFF:

1. Relatively easy to develop and implant
2. Useful where peripheral nerve is damaged or natural neural encoding is complex (e.g. joint angle)
3. Inherently electromechanically complex, leading to long term reliability problems
4. Must interact mechanically with living connective tissue

### IMPLANTABLE SENSORY NERVE ELECTRODES:

I. Whole Nerve Cuff - gross indicator of mixed sources
II. Single Unit Microelectrode - unstable indicator of a single random source
III. Multi-Microelectrode Array - complex multi-channel information requiring interpretation and tracking

### TRADE-OFF:

1. Requires sophisticated R & D
2. Requires surgery on nervous tissue
3. Requires complex signal conditioning and processing
4. Inherently electromechanically simple
5. Utilizes highly sophisticated natural transducers
6. Locatable in body core where electro-mechanically protected

# REFERENCES

1. **Peckham, P. H. and Mortimer, J. T.,** Restoration of hand function in the quadriplegic through electrical stimulation, in *Functional Electrical Stimulation*, Hambrecht, F. T. and Reswick, J. B., Eds., Marcel Dekker, New York, 1977.
2. **Vodovnik, L., Kralj, A., and Caldwell, C. W.,** Development of orthotic systems using functional electrical stimulation and myoelectric control, Final Report (Project 19-P-58391-F-01), Laboratory of Medical Electronics and Biocybernetics, University of Ljubljana, Yugoslavia, 1977.
3. **Sabine, C. L., Addison, R. G., and Fisher, H. K. J.,** A plastic tenodesis splint, *J. Bone Jt. Surg.,* 47A, 533, 1965.
4. **Donaldson, P. E. K.,** Experimental visual prosthesis, *IEEE Proc. (Great Britain),* 120, 281, 1973.

5. **Marks, W. B. and Loeb, G. E.**, Action currents, internodal potentials, and extracellular records of myelinated mammalian nerve fibers derived from node potentials, *Biophys. J.*, 16, 655, 1976.
6. **Stein, R. B., Nichols, T. R., Jhamandas, J., Davis, L. and Charles, D.**, Stable long-term recordings from cat peripheral nerves, *Brain Res.*, 128, 21, 1977.
7. **Strumwasser, F.**, Long-term recording from single neurons in brain of unrestrained mammals, *Science*, 127, 469, 1958.
8. **Salcman, M. and Bak, M. J.**, Design, fabrication and *in vivo* behavior of chronic recording intracortical microelectrodes. *IEEE Trans. Bio. Med. Electron*, 20., 253, 1973.
9. **Prochaszka, A., Westerman, R. A., Ziccone, S. P.**, Discharges of single hindlimb afferents in the freely moving cat, *J. Neurophysiol.*, 39, 1090, 1976.
10. **Loeb, G. E., Bak, M. J., and Duysens, J.**, Long-term unit recording from somatosensory neurons in the spinal ganglia of the freely walking cat, *Science*, 197, 1192, 1977.
11. **Schmidt, E. M., Bak, M. J., and McIntosh, J. S.**, Long-term chronic recording from cortical neurons, *Exp. Neurol.*, 52, 496, 1976.
12. **Watanabe, S., Ed.**, *Methodologies of Pattern Recognition*, Academic Press, New York, 1969.
13. **Walmsley, B., Hodgson, J. A., and Burke, R. E.**, The forces produced by hindlimb muscles in freely moving cats, presented at Soc. Neuroscience 7th Annu. Meet., Abstr. 896, 1977.
14. **Loeb, G. E., Bak, M. J., Salcman, M., and Schmidt, E. M.**, Parylene as a chronically stable, reproducible microelectrode insulator, *IEEE Bio. Med. Electron.*, 24, 121, 1977.
15. **Prochazka, A., Westerman, R. A., and Ziccone, S. P.**, Remote monitoring of muscle length and EMG in unrestrained cats, *Clin. Neurophysiol.*, 37, 649, 1974.
16. **Miller, S., van der Burg, J., and van der Meche, F. G. A.**, Coordination of movements of the hindlimbs and forelimbs in different forms of locomotion in normal and decerebrate cats, *Brain Res.*, 91, 217, 1975.
17. **Grigg, P. and Greenspan, B. J.**, Response of primate joint afferent neurons to mechanical stimulation of knee joint, *J. Neurophysiol.*, 40, 1, 1977.
18. **Nashold, B. S., Friedman, H., Grimes, J., and Avery, R.**, Electromicturition in the paraplegic: an electroneuralprosthesis to control voiding, in *Neural Organization and Its Relevance to Prosthetics*, Fields, W. S., Ed., Intercontinental Medical Book Corp., New York, 1973.
19. **Peckham, P. H., Mortimer, J. T., and Marsolais, E. B.**, Alteration in the force and fatiguability of skeletal muscle in quadriplegic humans following exercise induced by chronic electrical stimulation, *Clin. Orthop. Relat. Res.*, 114, 326, 1975.

# DISCUSSION

**Dr. Ernest P. McCutcheon:** I would like you to comment again on the relationship between EMG and muscle tension.

**Dr. Loeb:** There's no question that there is a correlation, more EMG usually means there is more tension. We have found it useful to make assumptions that when there is a bigger EMG, it is likely the muscle is producing more tension. Exactly how much more tension is very difficult to determine. If you're talking about step cycles and choose a particular time in the step cycle when the muscle is moving at a certain rate, and you look at one step cycle where there's more EMG than another, you probably can say there's more tension. However, if you look at the EMG alone, for instance, and compare a situation where the muscle is lengthening against its contraction, EMG is going to be very nonlinear. If you're looking at rapid modulations in EMG activity, then you're going to have problems with phase advance because the EMG activity leads the contraction by, perhaps, 50 msec.

**Dr. Sidney Wolfson:** Would you comment on what work has been done to indicate the duration of the effective or accurate recording from such electrode arrays in chronic implantation?

**Dr. Loeb:** The longest animal we have hooked up this way (simply because it suits our purposes physiologically) has given us 2 months of recordings from individual units. Typically, each electrode in a ganglion works. It records units, and it records one to three separable unit action potentials which are individually stable for, in our cases, several days. The longest have been 3 to 4 weeks from a single unit. Typically, it really isn't more than about 3 or 4 days. We've made no attempt to improve the anchoring of these electrodes, to reduce the stiffness of the wire so that it's more flaccid and is not subject to strains pushing it through the ganglion, or to examine how to minimize the foreign body reaction which is probably walling off the tip to some extent. We do know it is possible to record unit information in the motor cortex from this sort of floating electrode for over a year, not necessarily from the same neuron, however. We have not attempted to improve that stabilization because we really don't want to record the same unit for months. We would rather start with one unit, study it a couple of days, learn lots about it, and then go on to another unit. Otherwise, we have to do another cat.

**Dr. Wolfson:** Yes, but of course the total future of this approach depends upon being able to do this for very long periods of time.

**Dr. Loeb:** My guess is that it will be impossible to guarantee recording from a given nerve fiber for years. You will have to record from multiple nerve fibers, sample the population, and track the populations. They'll be stable for days, possibly weeks, but you're going to have to have a smart electronic system coupled to the electrodes to interpret the signals.

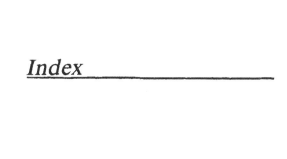

*Index*

# INDEX

Printed and bound by CPI Group (UK) Ltd, Croydon, CR0 4YY

22/10/2024

01777632-0016